Materials for Non-linear and Electro-optics 1989

Materials for Non-linear and Electro-optics 1989

Proceedings of the International Conference held in Cambridge, England, 4–7 July 1989

Edited by M H Lyons

Institute of Physics Conference Series Number 103
Institute of Physics, Bristol and New York

CODEN IPHSAC 103 1–252 (1989)

British Library Cataloguing in Publication Data

Materials for non-linear and electro-optics 1989.
   1. Electro-optical equipment. Materials
   I. Lyons, M.H.   II. Institute of Physics
   621.36

   ISBN 0-85498-059-8

Library of Congress Cataloging-in-Publication Data are available

Honorary Editor
   M H Lyons

Published under The Institute of Physics imprint by IOP Publishing Ltd
Techno House, Redcliffe Way, Bristol BS1 6NX, England
335 East 45th Street, New York, NY 10017-3483, USA

Printed in Great Britain by J W Arrowsmith Ltd, Bristol

Contents

Preface

This volume contains the invited and contributed papers presented at the International Conference on *Materials for Non-linear and Electro-optics* which was held at Girton College, Cambridge University between 4 and 7 July 1989. The conference was jointly sponsored by the Thin Films and Surfaces Group of The Institute of Physics (IOP) and the Industrial Physical Chemistry Group of The Royal Society of Chemistry.

The strong world-wide interest in materials with non-linear or electro-optic properties was demonstrated by the fact that half the 100 delegates attending the conference came from outside the UK. A total of 16 countries were represented. An underlying theme of this conference was that interest in these non-linear and electro-optic materials was driven by the demand for novel optic and electro-optic devices. Papers included in this volume reflect the very wide range of materials currently being investigated for device applications, ranging from inorganic crystalline solids to polymers and liquid crystals.

Each camera-ready paper submitted for publication in this volume was reviewed by at least one referee and I am very grateful to the following referees for their prompt and efficient work: S Allen, D J Ando, R T Bailey, F R Cruickshank, K J Harrison, G R Meredith, P Pantelis, D A Payne, M C Petty, K Skarp, W J Stewart, P A Thomas, R Whatmore, B S Wherrett and J Zyss.

The organization of a conference depends on the efforts of many people. I would therefore like to acknowledge the assistance given by staff at the IOP and IOP Publishing Ltd, and the help and encouragement given by my colleagues on the organizing committee: D J Ando, R T Bailey, P Pantelis, M C Petty, W J Stewart and R Whatmore. I also acknowledge the support given to the conference by IOP Publishing Ltd and BT&D Technologies Ltd.

M H Lyons
August 1989

Applications of nonlinear and electro-optic materials

I Bennion

Plessey Research Caswell Ltd., Caswell, Towcester, Northants, NN12 8EQ, U.K.

ABSTRACT: Applications of nonlinear and electro-optic materials are reviewed from the standpoint of optical or optoelectronic component functionality. Guided-wave electro-optic devices for functions including phase, amplitude, frequency and polarisation modulation, and switching, are reviewed; integrated optical combinations of these devices to yield components for applications in communications, signal processing, instrumentation and sensing are considered. Some applications of the second- and third-order nonlinear optical susceptibilities in all-optical guided-wave device formats are described. Applications of electro-optic materials in two-dimensional device arrays for optical interconnection and parallel processing, offering opportunities for new materials, are identified.

1.   INTRODUCTION

Applications for devices based on electro-optic and nonlinear optic materials can be identified in the fields of optical communications, signal processing, sensing, and instrumentation. This paper is intended to provide an introduction to the functions currently pursued by device designers for these applications. Waveguide formats have been adopted for the optical interactions for several reasons: (1) long interaction lengths can be maintained offering higher interaction efficiencies than are achievable using freely propagating laser light. (2) High guided-wave power densities can be achieved from moderate input optical powers due to the confinement imparted by small waveguide cross-sections, which is significant particularly for $\chi^{(2)}$ and $\chi^{(3)}$ nonlinear interactions. (3) Guided-wave devices can be combined to produce more complex functions in single, monolithic components by the techniques of *integrated optics*. Whilst device formats and applications based on materials used in non-guided-wave formats readily may be conceived, a far greater range of practical structures currently employs optical waveguides and, for this reason, we concentrate predominantly herein on guided-wave devices.

In this paper, we distinguish two categories of device according to whether operation is through either an electro-optic effect (linear or quadratic), whereby the interaction is induced by an applied electrical signal (Kaminow & Turner 1966), or a nonlinear optical effect ($\chi^{(2)}$ or $\chi^{(3)}$) whereby the interaction is entirely optical. Devices in the first category are substantially more mature and provide a wider range of component functions in guided-wave form at the present time; useful reviews are by Alferness (1981, 1982), Thylén (1988) and Syms (1988). Guided-wave nonlinear optics offers many potential future devices, and the field is well reviewed in Stegeman & Seaton (1985), Stegeman & Stolen (1988) and Stegeman et al (1988).

Current device interests in electro-optic and nonlinear optical materials do not lie exclusively in the guided-wave domain, however. There is considerable interest in the use of optical techniques to perform parallel processing or computational functions based on the use of electro-optic or nonlinear optical elements in two-dimensional arrays. It is extremely difficult to conceive such arrays utilising guided-wave technology and they generally use unguided interactions. This class of device, representing an opportunity for new materials with enhanced electro-optic or nonlinear optic coefficients, is also addressed in this paper.

2.   WAVEGUIDES AND MATERIALS

Many waveguide structures may be used in device implementations based on electro-optic or nonlinear optical materials some of which are illustrated in Figure 1. In each of the illustrated structures $n_2 > n_1$; in the strip-loaded case, n_3 may be greater or less than n_1, n_2. Selection of the appropriate waveguide type is dictated by the combination of material properties and processing methods available. The channel structure is created by masked diffusion, ion-exchange, ion implantation, or selective poling in organic materials: the best known example is Ti diffusion in $LiNbO_3$. The rib structure is produced by etching and is utilised with epitaxially grown layers such as III-V semiconductors. The strip-loaded structure is also produced by etching and finds application in GaAs/GaAlAs devices; it is also, however, a convenient way in which to create waveguide structures incorporating a layer of a nonlinear material (n_2) for characterisation purposes or where effective strip waveguide processing conditions have not been established.

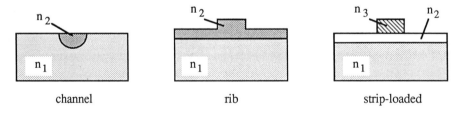

channel   rib   strip-loaded

Fig. 1. Optical waveguides.

Typical waveguide cross-sectional dimensions are in the range ~ 1-$50\mu m$ for single-mode operation. Smaller cross-sections are appropriate for minimising the nonlinear or electro-optic interaction volumes. However, devices are commonly required to be interfaced to optical fibres with mode diameters of ~ 8-$10\mu m$. Efficient, lensless power coupling can only then be achieved when the fibre and waveguide mode sizes are equal and larger waveguide dimensions are then necessary. Lens-coupling between fibre and a relatively smaller waveguide is, of course, possible at the expense of increased component complexity and severity of the alignment tolerance.

3. GUIDED-WAVE ELECTRO-OPTIC DEVICES

A light wave injected into a linear electro-optic medium undergoes a phase shift $\phi = kVL$ proportional to an applied voltage V, where L is the interaction length and k is a constant dependent on the material properties and the interaction geometry. In the case of a quadratic (Kerr) electro-optic medium, $\phi = KV^2L$.

The most elementary electro-optic device is, therefore, a phase shifter or phase modulator, shown in waveguide form in Figure 1 and consisting simply of a waveguide of length L and associated electrodes. Elements of this type have applications in many fibre optic systems both as self-contained devices and as constituent parts of more complex integrated optical circuits, as discussed in following Section 5.

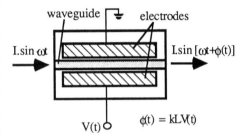

Fig. 2. Guided-wave electro-optic phase modulator

Alferness (1982) provides a detailed discussion of the design/performance trade-offs in phase modulators and derives the expression

$$V/\Delta f = \pi R \left(\frac{\varepsilon_{eff}}{n^3 r} \right) (p\lambda) \left(\frac{d}{\Gamma} . \xi \right) \tag{1}$$

for the voltage per unit bandwidth. In (1), λ is the wavelength, n the refractive index, ε_{eff} the dielectric constant, r the relevant electro-optic coefficient, Γ a factor with value between 0 and 1 describing the extent of the overlap between the interacting optical and electrical fields, d the electrode separation, R the terminating resistance, ξ a function of the electrode width and separation, and p a factor of order 1 which relates to the modulator type. The objective of good design is to minimise $V/\Delta f$ consistent with maintaining low optical attenuation through the device. From the viewpoint of the materials designer, this criterion favours materials with low dielectric constant, high electro-optic coefficients and high refractive index. For efficient device design, however, other factors must be considered. The factor Γ is dependent upon the electrode configuration. In the case of coplanar electrodes, appropriate to diffused waveguides in LiNbO₃, for example, Γ may be only 0.1~0.2 for a device operating at $\lambda=1.3\mu m$. In a GaAlAs/GaAs device, however, epitaxial growth procedures permit location of the electrodes directly below and above the waveguide giving $\Gamma{\sim}1$. Spin-coatable organic materials (e.g. Lytel et al 1988) are also compatible with this arrangement.

The achievable bandwidth is limited in the first instance by the RC time constant, τ_{RC}, of the modulator. In practice, there is a trade-off between minimising the electrode capacitance (e.g. by reducing the length L) and the increased voltage thereby required to achieve the required degree of modulation ($V_\pi \propto 1/L$). In order to overcome the bandwidth limit imposed by τ_{RC}, it is necessary to adopt technologically more demanding travelling-wave structures (Izutsu et al 1977, Becker 1984) in which the optical and electrical signals are injected at one end of the device and propagate codirectionally along its length achieving phase modulation by a cumulative interaction. In this case the bandwidth limiting factor is the velocity difference between the optical and electrical waves. Care must be exercised to minimise electrical losses in travelling-wave modulators. Guided-wave phase modulation has been demonstrated at frequencies up to 100GHz (Nees et al 1989).

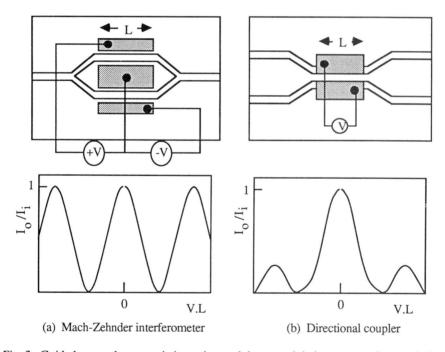

(a) Mach-Zehnder interferometer   (b) Directional coupler

Fig. 3. Guided-wave electro-optic intensity modulators and their response characteristics

There is greater practical interest in optical intensity modulation. Many device principles have been discussed (Alferness 1982) but two have received most attention: these are the Mach-Zehnder interferometer and the directional coupler, illustrated along with their transmission responses as functions of the product V.L in Figure 3. The principles of operation of these modulators are well documented (e.g. Alferness 1981, 1982) and are not repeated here: the issues addressed above in connection with phase modulators apply equally to these devices. Modulators of both types have been implemented in a range of electro-optic materials in both lumped element and travelling-wave electrode forms but are at their highest state of development in LiNbO$_3$ and GaAs at the present time. An extensive literature exists in the case of LiNbO$_3$ dealing with the many aspects of practical modulator design and fabrication: the papers by Thylén (1988) and Syms (1988) provide useful starting points. The papers by Walker (1987) and Wang & Lin (1988) serve a similar purpose for the III-V semiconductor materials which offer an alternative perspective on practical device implementation.

Design objectives for intensity modulators vary with the intended application. Generally, the goal, as for phase modulators, is to achieve some desired frequency performance with minimum expenditure of electrical drive power. The frequency range of interest is primarily from a few GHz upwards although applications certainly exist at lower frequencies. The bandwidth requirement often extends down to d.c. but many systems applications are perceived for r.f. bandpass devices. In these fractional bandwidth modulators, velocity mismatch compensation schemes (Alferness et al 1984a, Thylén and Djupsöbacka 1985, Erasme et al 1988), or electrical (Izutsu et al 1987) or optical (Stewart et al 1984) resonance techniques can be employed purposely to restrict the modulation bandwidth with some reduction in the drive power requirement. Broadband performance to 40GHz has been demonstrated in LiNbO$_3$ albeit using a velocity mismatch compensation scheme which yields considerable nonlinearity in the phase frequency response (Dolfi et al 1988). The wavelength range of interest is predominantly though by no means exclusively ~0.8-1.6μm compatible with compact solid-state laser sources.

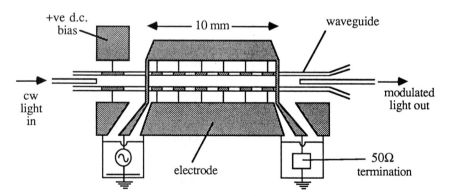

Fig. 4. High-speed travelling-wave Mach-Zehnder modulator with velocity-matching electrode configuration for GaAs/GaAlAs

A figure of merit, extracted from (1), which permits comparison of different modulator performances is $F = \Delta f.\lambda/V_\pi$. The highest reported value of F is 6.2GHz.µm.V^{-1} achieved with the GaAs/GaAlAs travelling-wave Mach-Zehnder modulator illustrated in Figure 4 (Walker, Bennion and Carter 1989a). This device embodies novel waveguide design principles first implemented in lumped electrode structures (Walker 1989). The travelling-wave device of Fig. 4 employs an electrical slow-wave electrode system to achieve near-velocity matching: bandwidth of 23GHz was achieved for 4.25V applied to the 50Ω structure with λ=1.15µm.

In addition to phase and amplitude modulation, many further functions are possible using guided-wave electro-optics. The simple phase modulator of Figure 3 can be used to provide optical frequency translation by an amount equal to the frequency of an applied sawtooth signal in the so-called serrodyne modulator utilised in interferometric sensors (Wong et al 1982). By appropriate choice of material orientation and electrode configuration, a single waveguide device can be made to perform polarisation mode conversion (McKenna and Reinhart 1976, Alferness 1980, Mariller and Papuchon 1985) which forms the basis for further functions, most notably, perhaps, polarisation control. Guided-wave modulators are well-suited as intracavity control elements for semiconductor lasers, for example, for mode-locking (Alferness et al 1984b) or frequency stabilisation and tuning (Korotky et al 1986, Heismann et al 1987). Reference to the literature cited herein provides many further device examples.

Guided-wave techniques permit high levels of functional complexity to be achieved by combining elements such as those described above into monolithic integrated optical circuits. The resulting devices are compact and rugged, eliminate the optical alignments and interfaces required by separate bulk component implementations, and offer, in principle, the cost advantages associated with planar device manufacture. We illustrate integrated electro-optical circuits with representative devices of current interest taken, respectively, from the fields of optical sensing, signal processing and, in the following section, communications.

Figure 5 shows an integrated optical device within a fibre optic gyroscope (FOG). The FOG is configured as a Sagnac interferometer: rotation of the fibre coil induces a phase shift between coherent light waves propagating in opposite directions around the coil which is a measure of the rotation rate (Ezekiel and Arditty 1982). The phase shift can be cancelled by an equal and opposite shift supplied by the integrated electro-optic modulators, in which case the applied voltage is a measure of the rotation rate. In this application, the electro-optic devices are combined with several purely passive elements and the single integrated optic component provides the necessary beamsplitting and recombining functions, a polariser and spatial filtering, as well as the modulators. Devices of this type, in several variants, are currently fabricated in LiNbO$_3$ and are perceived as one of the first major applications of integrated electro-optic technology.

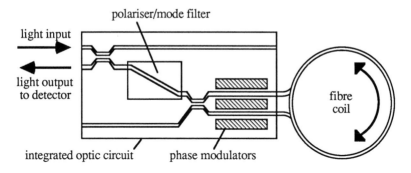

Fig. 5. Fibre optic gyroscope incorporating a multifunction integrated optic device

A device of somewhat greater complexity is illustrated in Figure 6 which shows an electro-optic 4-bit analogue-to-digital converter in a single-tapped Mach-Zehnder (STMZ) configuration. This device has been implemented in GaAs/GaAlAs (Walker et al 1989). Earlier devices using the well-known parallel Mach-Zehnder scheme were produced in LiNbO$_3$ (Becker et al 1984). In those structures, a series of interferometers with electrode lengths sequentially related by a factor 2 is driven by the same input voltage waveform. Taylor (1979) has shown that by thresholding the resultant outputs a binary representation of the applied waveform is derived. The STMZ device emulates with advantages the parallel interferometer arrangement (Walker and Bennion 1989). It uses one interferometer tapped along its length by directional couplers after 1, 2, 4 and 8 units of accumulated electrode length, accomplishing the same function as a parallel Mach-Zehnder structure with four separate interferometers but utilising less overall waveguide length and only ~50% of the electrode area. The lumped electrode elements in current devices restrict operating speed to ~1GHz; it is anticipated that this will be substantially increased in travelling-wave implementations.

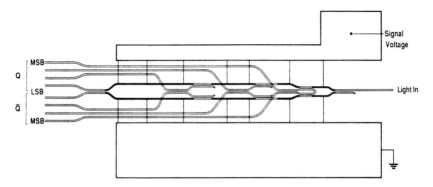

Fig. 6. 4-bit electro-optic analogue-to-digital converter

*   OPTICAL SWITCHING

In this section, we consider light wave switching by applied electrical signals via the electro-optic effect. (An alternative concept, that of light waves switched by optical signals, is considered in Section 7.) Interest in electro-optical switching stems not only from their high-speed, high-bandwidth capabilities but also from their transparency to the frequency content and modulation formats of the information they carry, an attribute not shared by electrical switches. Applications are envisaged in space-, time- and wavelength division switching, and in switching r.f. signals carried on fibre optic transmission lines. The directional coupler of Figure 3b is the most widely used switching element: applied voltage levels determine whether lightwaves injected at the inputs emerge in the same or opposite waveguides. Electro-optic switches are formed as interconnected arrays of elements (see Figure 7). The interconnection architecture within the chip is determined by performance and application constraints. In Ti:LiNbO$_3$, arrays as large as 8x8 have been reported in several different architectures (e.g. Granestrand et al 1986, Duthie and Wale 1988). Switch arrays are currently the most densely integrated of electro-optic structures. In LiNbO$_3$, the array size is now

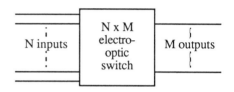

Fig. 7 Electro-optic switch

limited by the size of available substrates (75mm diameter) and efforts to expand switch sizes have led to the exploration of new device architectures (Duthie et al 1987) and chip interconnection techniques (Heidrich 1989). A detailed review of the status of guided-wave switch arrays is given by Alferness (1988).

In directional couplers wherein the TE and TM guided modes are influenced by unequal electro-optic coefficients, switch performance is highly polarisation dependent and this factor may limit the practical usefulness of these devices in many applications. To circumvent this limitation in LiNbO$_3$, a number of polarisation-independent switch structures have been devised (Kondo et al 1987, Alferness 1988, Granestrand and Thylén 1988, Nightingale et al 1989), usually at the cost of increased switching voltage, larger size, severe fabrication tolerances, or a combination of these. In order to maintain a high degree of crosstalk isolation, the directional coupler requires fairly precise voltage control and uniformity of the switching voltage required by all elements of an array is at least desirable, perhaps essential for practical utilisation. These requirements are addressed in a new form of electro-optic switch described by Silberberg et al (1987) wherein the two states exhibit saturation characteristics with increasingly positive- and negative-going voltages. This switch also demonstrates polarisation and wavelength insensitivity although both the voltage and the switching element size are relatively large.

5. ELECTRO-OPTIC MODULATOR ARRAYS

Applications can be identified for electro-optic modulators in the form of two-dimensional arrays, a format which cannot readily be achieved with guided-wave devices. Spatial light modulators (SLM), for example, are configured as arrays of liquid crystal modulator elements. Implementations of optical interconnects for electronic integrated circuits are envisaged which utilise arrays of optical modulators driven by these circuits. Figure 8 shows the cross-section through one element of a hybridised electro-optic modulator array compatible with these applications (Bennion et al 1989). The

modulator array in PLZT is flip-chip solder-bonded to a silicon IC. The individual modulator electrodes are driven from the IC via the solder bonds and associated metallisation. The device operates in reflective mode with light passing twice through the modulator apertures. In this case the PLZT device functions as a polarisation modulator and conversion to intensity modulation is accomplished by an analyser, not shown. For the SLM, the IC comprises a set of driving transistors. However, this

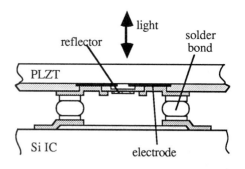

Fig. 8. Hybridised electro-optic modulator in PLZT solder-bonded to a silicon IC

hybridisation scheme is compatible with more general silicon circuits and, thus, provides a basis for the implementation of free-space chip-level optical interconnects.

The prototype of the device shown in Figure 8, an 8x8 array (Bennion et al 1989), utilises the quadratic electro-optic effect in PLZT. The technique, however, may be advantageously applied to other materials. The applications of modulator arrays on electrical ICs represents an opportunity for new materials. In particular, solution deposited organic materials may even remove the need for the solder-bonded connections.

6. GUIDED-WAVE NONLINEAR OPTICS

Device applications of second-order guided-wave nonlinear optics are concerned with second-harmonic generation (Stegeman and Seaton 1985) and parametric interactions (Sohler et al 1986). A major goal is efficient device structures for operation with semiconductor injection lasers. To date, $LiNbO_3$ has been predominant (Regener and Sohler 1988) although photorefractive instability caused by the high power densities at shorter wavelengths has necessitated exploration of enhanced strip waveguide fabrication techniques (Fejer et al 1986). An alternative approach utilises the crystalline nonlinear material in optical fibre form (Luh et al 1987). These applications represent a major opportunity for organic materials (Zyss 1985) in the forms of doped, poled polymers, single crystals, or Langmuir-Blodgett films if effective waveguide fabrication techniques can be established.

In third-order nonlinear optics, the $\chi^{(3)}$ coefficient gives rise to an intensity-dependent refractive index $n(I) = n_0 + n_2(I).I$ where n_0 is the low-intensity value of n, and I the intensity. This leads to an intensity-dependent propagation constant for light waves in $\chi^{(3)}$ media. It is easy to see that, in principle, all of the electro-optic device concepts discussed hitherto should have an all-optical counterpart wherein the role of the applied voltage is adopted by the power of the propagating light wave itself. The great attraction in the $\chi^{(3)}$ devices stems from their potential for very high speed modulation and switching, in the ps or sub-ps regimes. Applications are foreseen for devices which can generate and manipulate very fast optical pulses without electrical intervention: the device described by Lattes et al (1983) for optical logic operations amply demonstrates the principle.

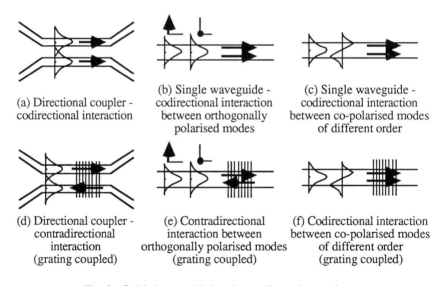

(a) Directional coupler - codirectional interaction

(b) Single waveguide - codirectional interaction between orthogonally polarised modes

(c) Single waveguide - codirectional interaction between co-polarised modes of different order

(d) Directional coupler - contradirectional interaction (grating coupled)

(e) Contradirectional interaction between orthogonally polarised modes (grating coupled)

(f) Codirectional interaction between co-polarised modes of different order (grating coupled)

Fig. 9. Guided-wave third-order nonlinear interactions

Figure 9 illustrates a number of guided-wave formats for $\chi^{(3)}$ interactions. These and other structures are described in Stegeman et al (1988). The nonlinear directional coupler shown in Fig. 9(a) was first discussed by Jensen (1982) and is the most-studied of the $\chi^{(3)}$ devices; theory predicts that it exhibits one of the sharpest switching transitions of the structures studied to date. Fig. 9(d)-(f) show applications of gratings as nonlinear mode coupling elements. Gratings, formed as surface relief structures or volume fringe patterns, are versatile structures in guided-wave optics generally (Yariv and Nakamura 1977): nonlinear versions take advantage of the high sensitivity of the Bragg condition to changes in propagation constant.

Detailed investigation of the materials requirements for useful $\chi^{(3)}$ devices reveals a fairly severe set of constraints that have to be met. In addition to possessing a suitable value of n_2, the response time of the nonlinearity must approach the sub-ps range, the absorption must be very low to avoid thermal effects and permit a reasonable optical throughput, and the material must be compatible with waveguide fabrication. To date, no ideal materials have been found: indeed, judged by one figure of merit (Stegeman et al 1988), SiO_2 could be claimed to be the best yet identified, which is borne out in part at least by nonlinear experiments on optical fibres. This is without doubt an area which will benefit from the emergence of new materials.

7. CONCLUSIONS

This paper has outlined some of existing and potential device applications for electro-optic and nonlinear optical materials, concentrating principally on the guided-wave format which offers the richest variety of applications. In the shorter term, more systems demands are likely to be met by electro-optic devices than those utilising nonlinear optical effects. In all cases, there is considerable scope for new materials. Improved materials need not necessarily have higher electro-optic or nonlinear coefficients, although these are always welcome; other physical parameters, ease and repeatability of processing, compatibility with waveguide design and fabrication techniques, damage thresholds and stability are equally important.

8. REFERENCES

Alferness R C 1980, *Appl Phys Lett* **36** 513.
Alferness R C 1981, *IEEE J Quantum Electron* **QE-17** 946.
Alferness R C 1982, *IEEE Trans Microwave Theory & Techn* **MTT-30** 1121.
Alferness R C 1988, *IEEE J Select Areas Commun* **6** 1117.
Alferness R C, Korotky S K and Marcatili E A 1984a, *IEEE J Quantum Electron* **QE-20** 301.
Alferness R C et al 1984b, *Appl Phys Lett* **45** 944.
Becker R A 1984, *Appl Phys Lett* **45** 1168.
Becker R A et al 1984, *Proc IEEE* **72** 802.
Bennion I, Goodwin M J, Groves-Kirkby C J and Parsons A D 1989, *Topical Meeting on Optical Computing,* Salt Lake City, Utah, USA, March 1989, Tech. Digest pp. 50-53.
Dolfi D W, Nazarathy M and Jungerman R L 1988, *Electron Lett* **24** 528.
Duthie P J and Wale M J 1988, *Electron Lett* **24** 594.
Duthie P J, Wale M J and Bennion I 1989, *Topical Meeting on Photonic Switching,* Salt Lake City, Utah, USA, March 1989, Tech. Digest pp. 174-176.
Erasme D et al 1988, *J Lightwave Technol* **6** 847.
Ezekiel S and Arditty H J (eds.) 1982, *Fibre-Optic Rotation Sensors and Related Technologies,* Springer-Verlag, Berlin.
Fejer M M, Digonnet M J F and Byer R L 1986, *Opt Lett* **11** 230.
Granestrand P et al 1986, *Electron Lett* **22** 816.
Granestrand P and Thylén L 1988, *Techn. Digest IGWO '88,* Santa Fe, USA, 244.
Heidrich H 1989, *7th International Conf. Integrated Optics and Optical Fibre Communication,* Kobe, Japan, July 1989, Techn. Digest, Vol 4, pp. 8-9.
Heismann F et al 1987, *Techn. Digest OFC/IOOC '87,* Reno, Nevada, USA, 149.
Izutsu M, Yamane Y and Sueta T 1977, *IEEE J Quantum Electron* **QE-13** 287.
Izutsu M et al 1987, *Techn. Digest OFC/IOOC '87,* Reno, Nevada, USA, 126.
Jensen S M 1982, *IEEE J Quantum Electron* **QE-18** 1580.
Kaminow I P and Turner E H 1966, *Proc IEEE* **54** 1374.

Kondo M et al 1987, *Electron Lett* **23** 1167.
Korotky S K et al 1986, *Appl Phys Lett* **49** 10.
Lattes A et al 1983, *IEEE J Quantum Electron* **QE-19** 1718.
Luh Y S 1987, *J Crystal Growth* **85** 264.
Lytel R et al 1988, *Proc SPIE* **971** 218.
McKenna J and Reinhart F K 1976, *J Appl Phys* **47** 2069.
Mariller C and Papuchon M 1985, *Proc 3rd European Conf Integrated Optics,* Berlin, 174.
Nees J, Williamson S and Mourou G 1989, *Appl Phys Lett* **54** 1962.
Nightingale J L, Vrhel J S and Salac T E 1989, *Integrated and Guided Wave Optics,* 1989
     Techn Dig Ser Vol 4 (Opt Soc Am, Washington D C) 10.
Regener R and Sohler W 1988, *J Opt Soc Am B* **5**, 267.
Silberberg Y, Perlmutter P and Baran J E 1987, *Appl Phys Lett* **51** 1230.
Sohler W et al 1986, *J Lightwave Technol* **4** 772.
Stegeman G I 1985, *J Appl Phys* **58** R57.
Stegeman G I and Stolen R H (eds.) 1988, *J Opt Soc Am B* **5**(2), February 1988, special
     issue on Nonlinear Guided-Wave Phenomena.
Stegeman G I et al 1988, *J Lightwave Technol* **6** 953.
Stewart W J, Bennion I and Goodwin M J 1984, *Phil Trans R Soc Lond* **A313** 401.
Syms R R A 1988, *Opt Quantum Electron* **20** 189.
Taylor H F 1979, *IEEE J Quantum Electron* **QE-15** 210.
Thylén L 1988, *J Lightwave Technol* **6** 847.
Thylén L and Djupsöbacka A 1985, *J Lightwave Technol* **3** 47.
Walker R G 1987, *J Lightwave Technol* **5** 1444.
Walker R G 1989, *Appl Phys Lett* **54** 1613.
Walker R G and Bennion I 1989, *Proc. 5th European Conference on Integrated Optics,* Paris,
     April 1989.
Walker R G, Bennion I and Carter A C 1989a, *7th International Conf. Integrated Optics and
     Optical Fibre Communication,* Kobe, Japan, July 1989, Techn. Digest, Vol 5, pp. 10-11.
Walker R G, Carter A C and Bennion I 1989b, *7th International Conf. Integrated Optics and
     Optical Fibre Communication,* Kobe, Japan, July 1989, Techn. Digest, Vol 4, pp. 78-79.
Wang S Y and Lin S H 1988, *J Lightwave Technol* **6** 758.
Wong K K, De La Rue R M and Wright S 1982, *Opt Lett* **7** 546.
Yariv A and Nakamura M 1977, *IEEE J Quantum Electron* **QE-13** 233.
Zyss J 1985, *J Mol Electron* **1** 25.

Inst. Phys. Conf. Ser. No 103: Part 1
Paper presented at Int. Conf. Materials for Non-linear and Electro-optics, Cambridge, 1989

13

Sol-gel processing of thin-layer dielectrics in lead-titanate-based systems

K. D. Budd and D. A. Payne

Department of Materials Science and Engineering, Materials Research
Laboratory and Beckman Institute,
University of Illinois at Urbana-Champaign, Urbana, Il 61801, USA

ABSTRACT: Thin-layer dielectrics were prepared by spin-casting methods. Details are given for the sol-gel process. Dense amorphous layers were formed at 300°C, and crystallization occurred above 450°C. The electric strengths were in excess of 20MV/m. Refractive index values were 2.4-2.6. Values of dielectric constant ranged from 30 for amorphous layers to 1700 for polycrystalline PLZT 7/65/35. PZT 53/47 was ferroelectric with a remnant polarization of 32.5 μC/cm^2. An electrooptic effect was demonstrated in a PLZT optical switch.

1. INTRODUCTION

This paper describes the preparation and properties of thin-layer dielectrics ($\leq$ 1 μm) obtained by sol-gel methods. A novel processing route is outlined based upon the partial polymerizable condensation of organometallic precursors. As such it represents a potentially upset technology in ceramic processing. The method avoids powders and their handling problems. Emphasis is placed on chemical synthesis, densification and crystallization at reduced temperatures. The purpose is to integrate dielectrics onto semiconductors. Materials include $PbTiO_3$, $Pb(Zr,Ti)O_3$ and $PbLa(Zr,Ti)O_3$. The materials are polar and have structure-property relations suitable for pyroelectric, piezoelectric, ferroelectric and electrooptic applications. They are sensors and responders, and transduce information. They respond in a logical sense to an external stimulus or stress (T, σ, E) and have memory. Logic and memory are prerequisites for intelligent systems. Smart ceramics will find increasing applications if they can be integrated directly with semiconductors for control and feedback functions. Thus the present interest in the reduced temperature processing of $PbTiO_3$-based materials by sol-gel methods.

2. BACKGROUND

Lead titanate is a tetragonal perovskite with a ferroelectric- paraelectric transition near 500°C. The c/a ratio is 1.06 at room temperature. Because of the large distortion the phase transformation tends to be disruptive on cooling from the high temperature cubic state. Shape and volume changes cannot be accommodated by twinning, and Suchicital et al. (1986) have shown that microcrack formation leads to brittle fracture and failure of the ceramic. This is unfortunate, since the calculated spontaneous polarization is 80 μC/cm^2. Use of pure PbTiO$_3$ devices has been severely limited by methods based upon conventional high temperature sintering practices. Again the interest in low temperature chemical processing routes.

Numerous derivatives to PbTiO$_3$ are used, due in part to a large degree of crystalline solubility with other members of the perovskite family of materials. The additives stabilize the structure by reducing the tetragonal distortion. Many dielectric devices are based upon modified lead titanate. A particularly useful class is based on compositions in the PbTiO$_3$-PbZrO$_3$ system. "PZT" is extensively used in piezoelectric applications. La-modified compositions, "PLZT", can be densified to optical transparency, and are used in electrooptic devices. This paper reports on the sol-gel processing of pure PbTiO$_3$, and on modified compositions in the PZT and PLZT systems.

Despite the variety of useful properties which are now attainable, a large number of exciting potential applications are as yet unrealized. Applications involving microelectronics are restricted because of size limitations, and there is a dire need to develop processing methods for the deposition of dielectric materials onto integrated circuits. Devices which could make use of these configurations, include: displays, image storage media, non-volatile ferroelectric gate FET's, optical shutters and modulators, are often limited by high operating voltages or high fabrication costs. Many of these problems could be solved by the development of a suitable processing method for the fabrication of complex thin-layers, including ferroelectric and/or high permittivity materials such as PLZT. To date, most PLZT thin-layers are prepared by sputtering or evaporation. While progress is made, a number of problems have been associated with these methods. Setup and fabrication costs can be high because of the need for complex vacuum systems. There is also difficulty with composition and thickness control, and degradation of insulation resistance in vacuum. Ease of composition adjustment is necessary for the optimization of properties in complex systems like PLZT.

3. SOL-GEL PROCESSING

In this paper, we report on a potentially simple and inexpensive method for the fabrication of thin-layers of PbTiO$_3$-based materials. The method is based on a sol-gel technique, in which the requisite metal atoms are mixed initially in homogeneous liquid solution. Thin-layers are formed by spreading a layer of solution onto a desired substrate, followed by transformation of the liquid into a rigid gel-like structure. The gel layer is then converted into a dense inorganic dielectric layer by low temperature heat treatment (Budd, 1986a).

Sol-gel processing offers several potential advantages for thin-layer fabrication. It shares many of the advantages now commonly associated with chemical methods of ceramic processing, such as the potential for high purity, molecular mixing, chemical homogeneity, control of stoichiometry, and reduction of densification temperatures. Sol-gel processing is unique in that it also offers diverse forming opportunities. Powders, bulk monolithic parts, composites, fibers, and thin-layers can all be formed. Processing of thin-layers is particularly attractive since only small amounts of precursor are needed, and layers can be formed by simply dip-coating or spin-casting methods. The latter is amenable to planar semiconductor technology. The need to process and consolidate powders by conventional ceramic processing methods is thus avoided.

The phrase "sol-gel processing" is commonly applied to a processing method which involves a transition from a fluid to a gel-like state. Polymeric gels are formed from organometallics and require molecular species to be at least trifunctional, and polymerize into large three dimensional networks without precipitation. A variety of precursors have been used to form polymeric gels. By far the most versatile and widely used class of organometallic precursors at the present time are metal alkoxides. Metal alkoxides, e.g., $M(OR)_n$, consist of a metal atom, M, bonded through oxygen to one or more alkyl groups, R, where n is the valence of the cation. The gels reported on in this paper were formed from alkoxide containing precursors. Gelation in alkoxide-based polymeric systems consists of two principal reactions: (i) hydrolysis and (ii) polycondensation. Species which hydrolyze rapidly, polymerize first.

The hydrolysis of metal alkoxides involves the replacement of one or more alkoxy group by hydroxyl groups. Alcohol is a product of the reaction. Full hydrolysis would proceed by

$$M(OR)_n + n\,H_2O \quad \rightarrow\ M(OH)_n + nROH \qquad\qquad , \qquad (1)$$

and in this case, the hydrocarbon content would be preferentially removed by evaporation of the alcohol, and not by pyrolysis or binder-burn out as in conventional ceramic processing. Most spin-casting formulations use partially hydrolyzed solutions. Oxide formation involves condensation of hydroxy groups by (i) dehydration and/or (ii) dealcoholation reactions, e.g.,

$$2M(OR)_{n-1}\,OH \quad \rightarrow \quad M_2O(OR)_{2n-2} + H_2O \qquad , \qquad (2)$$

$$M(OR)_n + M(OR)_{n-1}\,OH \quad \rightarrow \quad M_2O(OR)_{2n-2} + ROH \qquad . \qquad (3)$$

As stated, gelation in alkoxide-based polymeric systems consists of two principal reactions, (i) hydrolysis and (ii) polycondensation. Successful gel formation requires the retention of solubility during polycondensation. Metal alkoxides hydrolyze vigorously with water, which can result in the formation of insoluble hydroxides, oxides and oxohydroxides. Therefore, alcohols are commonly used as solvents for alkoxides. Solubility is maintained by retaining sufficient amounts of OR groups, and/or forming non-fully condensed solvated structures. Therefore excess alcohol is used with insufficient water for complete alkoxy exchange.

Heterogeneous-M_1-O-M_2- linkage may preferentially be formed if the least reactive species are prehydrolyzed first, followed by additions of the more reactive alkoxide. Also, controlled condensation may be more reliably achieved if different functional groups, which are not capable of self- condensation, are associated with each cation. For example, the heterogeneous reaction of an acetate with an alkoxide, as indicated later in this paper, could give good control of homogeneity, especially in the absence of other reactions:

$$M\text{-}OR + Ac\text{-}M_2\text{-} \rightarrow \text{-}M_1\text{-}O\text{-}M_2\text{-} + RAc \qquad . \qquad (4)$$

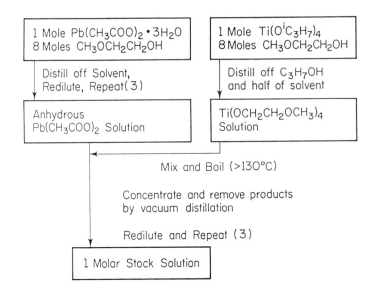

Fig. 1. Preparation of Pb,Ti-methoxyethoxide precursor solution

Additions of (i) acids or (ii) bases have been shown to promote (i) more linear or (ii) cross-linked networks, respectively (Dey et al. 1987). The physical properties of acid or base modified gels are a direct result of the type of networks which form, i.e., are dependent on the linear or more condensed structures. Clearly, the structure of polymeric gels is affected by solution chemistry, and the physical properties are correspondingly affected. The gel-to-ceramic conversion, which takes place upon heat treatment of a dried gel, is a complex process, and which is greatly dependent on gel structure. Driving forces for low temperature densification include a large available surface free area ($>150 m^2/g$) and free energy, a fine pore size distribution (2nm) and a large free volume (0.1 cc/g). These factors are influenced and controlled through manipulation of the gel structure by chemical and physical methods (Schwartz et al,1989).

4. EXPERIMENTAL

Details of the synthesis procedure have been reported elsewhere (Budd, et al. 1985, 1986b). Briefly, the precursors were Ti-isopropoxide and lead acetate (with Zr-n-propoxide and La-isopropoxide) in methoxyethanol. Separate acetate and alkoxide solutions were prepared using eight moles of methoxyethanol per mole of metal. Preparation conditions were under dry nitrogen using Schlenck apparatus. Figure 1 illustrates the synthesis procedure. The solutions were dehydrated by successive distillations, until the distillates were anhydrous to Karl Fisher titrations, and alcohol exchange reactions had gone to completion (e.g., Ti-methoxyethoxide). Stock solutions were formed by combining the alkoxide and acetate solutions at 100°C, followed by solvent and product distillation up to a boiling point of 135°C. The concentration was approximately 2.2-2.4 molar at this stage. The solution was later cooled to below 50°C, and three successive vacuum distillations were carried out. A maximum concentration of three molar

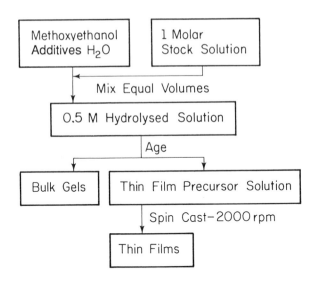

Fig. 2. Preparation of PbTiO$_3$ layers

was obtained, followed by redilution in methoxyethanol. The solution was then cooled, rediluted to one molar concentration, filtered and stored in a dry box.

Thin-layers were deposited from 0.5 molar PbTiO$_3$ solution. Figure 2 indicates that 0.5 molar solutions were formed by combining equal volumes of the one molar stock solution with a solution containing methoxyethanol, water, nitric acid or ammonium hydroxide. The time to gelation was principally controlled by the water content, and has been reported by Budd et al. (1986b). Thin-layers were deposited on silicon, platinized silicon, platinum and a variety of ceramic substrates, by spin-casting at 1000-4000 rpm on a photoresist spinner. The time was 30 seconds, slightly longer than the time required for interference colors to stop changing. Prior to spin-casting, the substrates were degreased with trichloroethylene and/or acetone, and rinsed with isopropanol and methoxyethanol. Silicon wafers were etched prior to cleaning with hydrogen peroxide and buffered HF solution. Platinum substrates were polished using a series of diamond pastes, followed by 0.3 micron and 0.05 micron alumina on a polishing microcloth. Precursor solutions were syringed through in-line filters directly onto the substrates in normal and clean-room conditions.

Thin layers were dried at 70°C for five minutes and stored in unsealed plastic boxes prior to heat treatment. A microprocessor-controlled electric furnace was used for heat treatment. Layer thickness and refractive index were determined on an ellipsometer. Excellent agreement was observed with the expected interference color. Densities were determined from the weight, thickness and area of coating on 2 inch diameter Si wafers. This information provided a calibration between density and refractive index as illustrated in Figure 3. Densification measurements were made after successive heat treatments and ellipsometer measurements. Crystallization behavior was determined by DTA/DSC thermal analysis and x-ray diffraction measurements. Electrical data were obtained as follows.

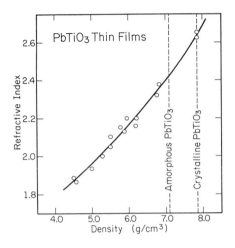

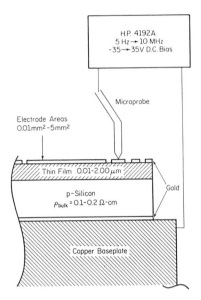

Fig. 3. Refractive index-density behavior

Fig. 4. Electrical measurements

Dielectric constant, loss tangent, electrical resistivity, and dielectric strength values were determined for amorphous and crystalline films deposited on platinum and silicon substrates. A parallel-plate capacitor geometry was used and is illustrated in Figure 4. Electrodes were patterned by evaporation through screens. Electrode areas ranged from 0.01 mm^2 to greater than 1 mm^2. The backs of the silicon wafers were coated to improve electrical contact.

Dielectric constant and loss tangent were measured with a Hewlett-Packard 4192 impedance analyzer, interfaced with an H.P. 9826 computer. This allowed point by point correction for finite closed circuit impedance and open circuit conductance. Frequency, oscillation level, and DC bias scans ranged from 1 kHz to 1 MHz; 5 mV to 1 V; and -30 V to 30 V; respectively. Generally, several electrodes (with areas spanning 1 to 2 orders of magnitude) were contacted to obtain dielectric constant values. Proper scaling of capacitance with area was taken to be an indication of an absence of probe-metal, or silicon-metal, contact capacitance problems. DC resistivities and dielectric strengths were calculated from current-voltage characteristics measured with an H.P. 4140B picoameter and DC voltage source. A ramp rate of 1 V per second was used. Again, a point-by-point correction for open circuit leakage was used.

Ferroelectricity was verified by polarization reversal measurements on a Sawyer-Tower bridge.

5. RESULTS

Information concerning chemical aspects of sol-gel synthesis and crystallization behavior of ceramic microstructures were reported by Budd et al. (1985, 1986b). The layers densified around 300°C and crystallized above 450°C. Emphasis is now placed on the optical and electrical properties.

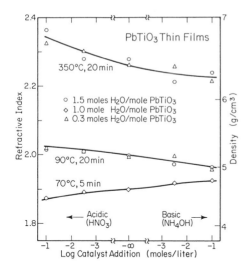

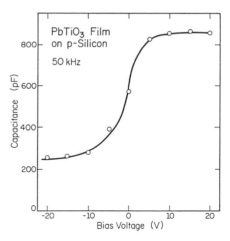

Fig. 5. Refractive index for amorphous PbTiO₃

Fig. 6. PbTiO₃ capacitor on Si

Previous discussion outlined the effect of acid and base additions on the development of polymeric networks and gel structure. Figure 5 illustrates the effect of additives on refractive index for PbTiO₃ thin-layers. The refractive index for dense (350°C) amorphous layers increased monotonically with the acidity of the precursor solutions. This has been attributed to the ease of visco-elastic rearrangement and densification for linear type structures (Budd, 1986a). The high densities were up to 92% of single crystal value. Since amorphous structures have additional free volume, nearly pore-free amorphous layers were formed at 350°C. The layers were transparent and could withstand 20 volts/μm i.e., 20 MV/m electric strength. Air dried or mildly dried (70°C, 5 minutes) layers exhibited the opposite trend in refractive index with acidity. This anomalous trend was attributed to residual organic material in the fine structures. Longer heat treatment (20 minutes) at a slightly higher temperature (90°C) was sufficient to reverse the trend.

Composition was by far the most important variable affecting crystallization temperature. For compositions studied, crystallization temperature increased with increasing Zr content. PbTiO₃, PZT 50/50 and PbZrO₃ crystallized at about 450°C, 500°C, and 550°C, respectively. The temperatures are below the calcination temperatures used in conventional mixed-oxide processing, and well below normal sintering or hot-pressing temperatures (>1000°C). The polycrystalline layers were dense and with a fine grain structure (≤0.1 μm).

The test-structure for thin-layers on Si was a MOS device. p-type Si, with an acceptor concentration greater than 10^{18} cm⁻³ was used. A reverse bias caused depletion of holes from the semiconductor-capacitor interface. The depletion layer developed a surface capacitance in series with the dielectric layer. Because of the additional series capacitor (which dilutes the total capacitance) the computed dielectric constant (K) for the dielectric layer would appear to be

Table 1: Dielectric Properties of Gel-Derived Thin-Layers
(50 KHz, 25°C, 1 V AC)

Composition	Crystallinity	Dielectric Constant	Loss Factor
PbTiO$_3$	amorphous	28 - 42	0.003 - 0.010
PbTiO$_3$	crystalline	155 - 185	0.006 - 0.030
PZT 53/47	amorphous	35	0.005 - 0.010
PZT 53/47	crystalline	300 - 1200	0.020 - 0.070
PZT 10/65/35	amorphous	35	0.005 - 0.010
PZT 10/65/35	crystalline	300 - 1700	0.020 - 0.090

less than normal. The effect would be more pronounced for layers with larger capacitances i.e., (thin high K-layers). Figure 6 illustrates the C-V characteristics for PbTiO$_3$ on silicon, which are consistent with a metal oxide structure on silicon. Hysteresis was sometimes observed in the C-V plots, probably due to incipient ferroelectric behavior. One to two volts forward bias were used for the dielectric constant measurements on Si. A summary is given in Table 1.

In general, the dielectric properties were good. Amorphous films had dielectric constant values ranging from 30-40, with low loss tangents less than 0.01, consistent with values reported by Takashige (1981) for dense, melt-quenched PbTiO$_3$ glasses. Polycrystalline PbTiO$_3$ films had K's ranging from 155 to 180, only slightly less than values compiled for bulk ceramic materials. Loss tangents ranged from 0.005 to 0.030. PZT and PLZT thin-layers had significantly higher dielectric constants, consistent with literature values for bulk materials.   The highest K's were obtained for compositions near the morphotropic phase boundary, as expected. Relative permittivities or dielectric constant values of up to 1700, for 1 micron layer of PLZT 10/65/35, and as high as 1200, for PZT 53/47, were measured.   The loss tangents of high-K compositions were larger than those for PbTiO$_3$ thin-layers.

Figure 7 illustrates the dielectric constant values (40-160) for a PbTiO$_3$ thin-layer as a function of heat treatment temperature. The sharp increase in dielectric constant near 500°C corresponded to the crystallization temperature for the material. For fully crystallized thin-layers, very little change in dielectric constant was observed with increasing heat-treatment temperature (500-700°C) although the grain-size increased from 100 to 1000 Å. The electrical resistivities were between 10^{11} and 10^{12} Ω -cm at 25°C.

PZT 53/47 and PLZT 10/65/35 had higher values of dielectric constant. Figure 8 illustrates the increase in dielectric constant as a function of heat treatment

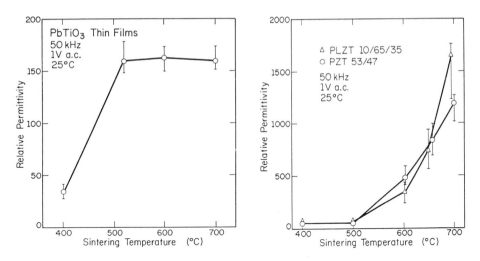

Fig. 7. PbTiO$_3$ dielectric   Fig. 8. PZT and PLZT dielectrics

conditions. Values increased for temperatures above the crystallization temperature (~550°C) as the grain size increased (100 - 1000Å). Ferroelectric properties were determined. Figure 9 illustrates a typical hysteresis loop for PZT 53/47. The squareness indicates the applicability for devices based upon switching. A remnant polarization of 32.5 µC/cm^2 was obtained for a 1 µm dielectric layer, and the coercive voltage was ±3V. It is clear that sol-gel processing offers a novel fabrication method for the deposition of non-linear dielectrics on substrates including semiconductors. A non-linear electrooptic effect has been demonstrated (Merkelo, 1988). Figure 10 illustrates the modulation of light from an end-fired waveguide. The waveguide was 1cm long

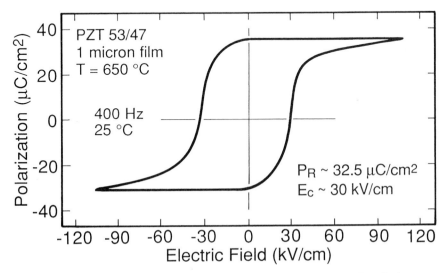

Fig. 9. Polarization-reversal for a sol-gel derived PZT dielectric layer

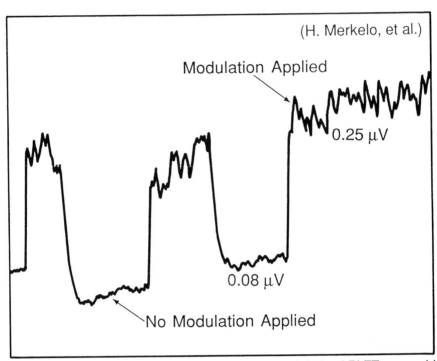

Fig. 10. Electro-optic modulation of light in a sol-gel derived PLZT waveguide

with transverse electrodes applied along the length, 200 μm apart. A chopped polarized He/Ne beam was used, and the output detected beyond an analyzer. Transmission was not detected between crossed polarizers, but observed after a transverse field had been applied. Figure 10 illustrates the transmission characteristics, as the voltage was modulated. A potential application for sol-gel thin layers would be in optical connectors.

ACKNOWLEDGEMENTS

The authors gratefully acknowledge support from an ONR-ASEE Fellowship and from US DOE DMR-DE-ACO2-76ER01198. Technical assistance from G. Manuel and A. Yu is recognized. Collaboration with S. K. Dey and H. Merkelo is gratefully acknowledged.

REFERENCES

Budd K D, Dey S K and Payne D A 1985 *Brit. Cer. Proc.* **36** 107
Budd K D 1986a *Ph.D. Thesis*, University of Illinois
Budd K D, Dey S K and Payne D A 1986 *Better Ceramics Through Chemistry II*
    ed C J Brinker et al. (Pittsburgh, MRS) 711-6
Dey S K, Budd K D and Payne D A 1987 *J. Am. Cer. Soc.* **70** 295
Merkelo H 1988 *Personal communication*
Schwartz R W, Payne and Holland A J 1989 *Proc. Cer. Powder Sci. II*, in press
Suchicital C T A, Payne D A and Grandin de l'Eprevier A 1986 *Proc. VI IEEE*
    ISAF pp 465-8
Takashige M 1980 *Jpn. J. Appl. Phys.* **19** L555

Inst. Phys. Conf. Ser. No 103: Part 1   23
Paper presented at Int. Conf. Materials for Non-linear and Electro-optics, Cambridge, 1989

High speed optical switches using sputtered PLZT thin films

H. Higashino, H. Adachi, K. Setsune, and K. Wasa

Matsushita Electric Industrial Co., Ltd., Moriguchi, Osaka 570, Japan

ABSTRACT: Thin film technology of PLZT has been developed in the past ten years and its applications have also been studied, especially in the field of optical switches. In this paper, the fabrication technology of PLZT thin films using RF-magnetron sputtering, and total internal reflection (TIR)-type and interference-type optical switches are described. These switches exhibit a low driving voltage of 4.7 V and a high operation frequency of 2 GHz, and they have the possibility of higher frequency operation up to over 80 GHz by using travelling wave electrodes.

1. INTRODUCTION

For integrated optics, thin films of $Pb_{1-x/100}La_{x/100}(Zr_{y/100}Ti_{z/100})_{1-x/400}O_3$ [PLZT(x/y/z)] are one of the most promising materials, because the PLZT thin films are expected to have a higher electro-optic (E–O) effect than $LiNbO_3$. The PLZT ceramic developed by Haertling et al(1971) is well known as a transparent ferroelectric material with a strong E–O effect. Ishida et al(1977) have shown that polycrystalline PLZT films deposited on Pt substrates by RF-diode sputtering exhibit a nonlinear E–O effect. Crystal structures and optical properties of epitaxial films grown on sapphire, MgO, and $SrTiO_3$ substrates have been reported by Ishida et al(1978) and Okuyama et al(1980). However, it was still difficult to obtain PLZT thin films of optical quality having high E–O effects. Recently we have succeeded in the fabrication of PLZT thin films on sapphire substrates(Adachi et al 1983a) and have tried to replace Ti-diffused $LiNbO_3$ waveguides with PLZT thin film waveguides in the $LiNbO_3$ total internal reflection (TIR)-type optical switches reported by Tsai et al(1978). We have refined the TIR-type switches(Wasa et al 1984 and Higashino et al 1985) to interference-type switches and demonstrated optical signal transmission using the interference-type switches(Higashino et al 1987). In this paper, we introduce an outline of the technologies of PLZT thin films and their optical switches.

2. SPUTTERED PLZT THIN FILM

In general, it is believed that high quality PLZT thin films of optical grade are difficult to obtain with high reproducibility.   One reason is the large difference in the vapor pressure of its constituent atoms, Pb, La, Zr, and Ti. A second reason is that the undesired pyrochlore structure often appears instead of the perovskite structure(Nakagawa et al 1979). Also, another reason is the formation of opaque PLZT films due to a growth of crystallites. Epitaxially grown single crystal films are more promising than polycrystalline films.

2.1 FABRICATION

Suitable substrates for epitaxial growth of PLZT thin films are reported as MgO and $SrTiO_3$(Okuyama et al 1980). Sapphire is also useful for the substrate due to its high thermal stability and the availability of an epitaxial grade. With respect to the crystal orientation, the c-plane of sapphire is most suitable for epitaxial growth. Its atomic configurations are shown in Fig. 1(Adachi et al 1986). The oxygen configuration shows good matching. The average distances of oxygen atoms are 2.75 Å for sapphire and 2.8 Å for PLZT. The lattice mismatch is about 2 percent. Their epitaxial relations are as follows:

(111)PLZT//(0001)sapphire
$[1\bar{1}0]$PLZT//$[10\bar{1}0]$sapphire   (1)

The composition of the PLZT thin film strongly affects its E-O properties. High E-O effects were observed at a composition near PLZT(9/65/35). However, the composition of Zr reduced the transparency of the sputtered films. For an optical waveguide, the simple composition PLZT(x/0/100) is recommended. Fig. 2 shows the variation of the E-O effect versus La content in the sputtered PLZT(x/0/100) system. The E-O effect in the plane was estimated by measuring the birefringence shift as shown in Fig. 3 of the paper by Adachi et al (1983a). The maximum E-O effect was observed at a La concentration of 28. Typical sputtering conditions are listed in Table 1.

Table 1. Sputtering conditions

Target	PLZT(28/0/100) powder
Substrate	c-plane sapphire
Target diameter	100 mm
Target-substrate spacing	35 mm
RF power	200 W
Sputtering gas	Ar(60%)+O_2(40%)
Gas pressure	0.5 Pa
Substrate temperature	580°C
Growth rate	70 ~ 80 Å/min

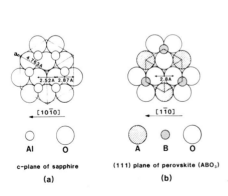

Al   O
c-plane of sapphire
(a)

A   B   O
(111) plane of perovskite (ABO₃)
(b)

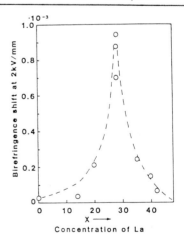

Fig. 1. Crystal structure of (111) plane of perovskite and c-plane of sapphire.

Fig. 2. Variation of the E-O effects vs. concentration of La in the target composition

2.2 FILM PROPERTIES

The composition of the films sputtered from the PLZT(28/0/100) target was PLZT(20/0/100). The sputtered films from the PLZT(28/0/100) target showed a Curie temperature of 150°C which was slightly higher than that of the PLZT(28/0/100) ceramics(Adachi et al 1983b). This difference was understood if the concentrations of La in the sputtered films were reduced to 20 percent or less. The PLZT(20/0/100) system shows a tetragonal structure with a slim-loop ferroelectric property. From the birefringence shift measurement, the quadratic E-O coefficient was derived to be 0.2 to $0.8 \times 10^{-16} (m/V)^2$ (Adachi et al 1983a). The E-O effect in TE_0 mode was about one order of magnitude greater than that in TM_0 mode(Kawaguchi et al 1984). These birefringence shifts were several times larger than for the $LiNbO_3$ single crystal under an applied field of 2 kV/mm and scarcely changed for any electric field direction in the (111) plane. The propagation loss of the waveguides was 5 dB/cm at an optical wavelength of 1.06 μm(Wasa et al 1984).

3. TOTAL INTERNAL REFLECTION-TYPE OPTICAL SWITCH

One kind of optical switch using PLZT thin films was the total internal reflection (TIR)-type optical switch. TIR-type optical switches using $LiNbO_3$ single crystals were first reported by Tsai et al(1978). The TIR-type optical switch needs a stronger E-O effect than the directional-coupler-type switches(Hammer 1975) but it has an advantage over them in high speed operation, because it has a short interaction length between an optical wave and a driving electric field. Wasa et al(1984) and Higashino et al(1985) have applied this structure to PLZT thin film optical switches.

3.1 STRUCTURES AND FABRICATIONS

Fig. 3 shows a typical configuration of type I of the TIR switch reported by Wasa et al(1984). The ridge waveguides were formed on the sputtered PLZT thin film on the sapphire substrate described above. The maximum thickness of the PLZT thin films was 350 nm and the quadratic E-O coefficient was $0.2 \times 10^{-16} (m/V)^2$ at 633 nm. The step height of the ridges was 50 nm. The width of the waveguides was 20 μm and they were used in transversal multimode. The intersecting angle was 2°. A pair of parallel metal electrodes of 4 μm separation and 1.7 mm in length were deposited at the center of the intersecting region. A buffer layer of Ta_2O_5 thin film about 150 nm thick was inserted between the PLZT thin film and the metal electrodes. The ridge waveguides were made by the conventional Ar ion beam etching method. The Ta_2O_5 buffer layer was deposited by sputtering from a Ta cathode in a mixed gas of Ar and O_2. The refractive index was 2.1 at 633nm.

The structure of type II of the TIR switch is shown in Fig. 4 as reported by Higashino et al(1985). In the switch, a strip-loaded type waveguide structure was adopted. The construction of the intersecting waveguide is shown in Fig. 5. The strips were made of Ta_2O_5 with a thickness of 20 nm and a refractive index of 2.09 at 633 nm. They were fabricated by the RF magnetron sputtering method and a lift-off technique. The width of the strips was 10 μm and the intersecting angle was 2°. The height of the strip in the crossing area was double that in the straight area. Therefore, the effective refractive index in the crossing area becomes double that in the straight area, Δn. The buffer layer was a mixture of Ta_2O_5 and Al_2O_3 with a refractive index of 1.93 and a thickness of about 150 nm. This was deposited by an RF-planar magnetron co-sputtering method with a Ta metal target and Al_2O_3 ceramic plates. The PLZT thin film had a quadratic E-O coefficient of $0.8 \times 10^{-16} (m/V)^2$ at 633 nm. Parallel metal electrodes of 4 μm separation, 0.9 mm in length and 1 mm in width were fabricated over the buffer layer.

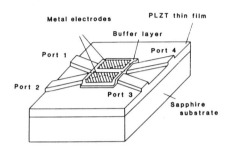

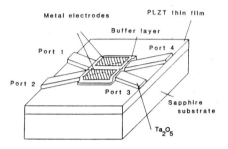

Fig. 3. Construction of TIR optical switch (type I).

Fig. 4. Construction of TIR optical switch (type II).

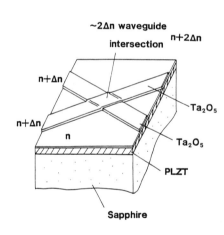

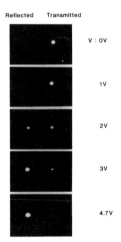

Fig. 5. Construction of the waveguide (type II).

Fig. 6. Near field pattern of the output waveguides in the TIR switch (type I).

3.2 SWITCHING OPERATIONS

The TIR-type optical switch can operate basically with multimode waveguides. In the absence of applied voltage between the parallel electrodes, the incident guided-light beam will propagate from port 1 to port 3 in Fig. 3 or Fig. 4 without any disturbance because the buffer layer prevents the metal electrodes from interfering with the guided light. When a bias voltage is applied, the refractive index of the PLZT thin film under the gap is reduced due to electro-optic effects. The incident guided-light beam from port 1 will be reflected by the reduced refractive index region and will pass into port 4. The reflection is similar to the optical total reflection in a 2-D material. In the TIR switch, the critical angle θ_c is given by

$$\theta_c = \sin^{-1}[1 - 0.5 N_{eff}^2 R(V/d)^2] \qquad (2)$$

where N_{eff} is the effective refractive index in the intersecting waveguides, R is the quadratic electro-optic coefficient, V is the applied voltage across the parallel electrode, and d is the spacing between the parallel electrodes.

Fig. 6 shows the near field pattern of the output waveguides of a type I switch for ports 3 and 4(Wasa et al 1984). The transmitted and reflected light beams observed through the GaP prism were switched by applying a driving voltage of 4.7 V under the superposition of 4.5-V dc bias. The dc-bias is effective not only for setting the birefringence shift at the optical threshold point but also for the cancellation of residual spontaneous polarization in the PLZT thin film. The cross talk measured at ports 3 and 4 was −13 and −12 dB, respectively.

In the measurement of switching, the guided-light beams in TE_0 mode were coupled in and out through the GaP prisms. However, the TE_0 mode will be partially converted into a higher order mode or TM mode in the PLZT thin films. The measured insertion loss of the TIR switches at their channel waveguides was 7 dB at 633 nm; this was mainly due to the propagation loss of the slab waveguide of 15 dB/cm at 633 nm. However, the insertion loss will be reduced to 2 dB or less at a wavelength of light longer than 1 μm, since the propagation loss was measured as less than 5 dB/cm at 1.3 μm.

In type II, the propagation losses of the PLZT slab waveguide in TE_0 mode were 14 and 6 dB/cm at 633 nm and 1.06 μm, respectively(Higashino et al 1985). The cross talk of the switch was reported as less than −18 dB before poling of the PLZT thin film.

3.3 HIGH SPEED OPERATION

The TIR-type optical switches have a potential for high speed operation as mentioned above. High speed modulation experiments were carried out by Wasa et al(1984) and Higashino et al(1985).

For type I, the measured capacitance between driving electrodes was less than 1 pF and a dielectric dispersion was scarcely observed at a frequency up to 1 GHz. The time constant of the device became 50 psec. In this device, operation at 1 GHz was carried out(Wasa et al 1984).

Over 1 GHz, operation of a TIR-type optical switch was reported for type II by Higashino et al(1985). High speed modulation of the switch and an optical transmission experiment was carried out using a laser diode (LD) at 830 nm. The guided wave was excited by the LD through the end-butt coupling and switched light was coupled out through a GaP prism and a graded refractive index (GRIN) lens to a multimode optical fiber of 50 μm core diameter. The microwave signal was fed through a semi-rigid cable to the electrodes with a matching terminator. The output optical signal was detected with an avalanche photo diode (APD). The frequency response was nearly flat up to 2 GHz when the applied signal level was 24 dBm (10 V_{p-p}). Fig. 7 shows the wave forms of the APD signal. The experimental frequency of 2 GHz was the instrumental limit.

An estimation of the operation bandwidth is possible by obtaining some data of phase mismatch between the optical guided wave and the microwave. To measure the effective dielectric constant of the switch in microwaves, the return loss of lumped constant asymmetric electrodes, which had a width of 0.6 mm, a length of 1 mm and a separation of 4 μm, was measured(Higashino et al 1985). The measured 3 dB cut-off frequency of the return loss was 7.05 GHz. This means that the switch can operate up to 7 GHz, assuming that the E-O coefficient does not decrease. From this result, the effective capacitance of 0.904 pF, which included a stray capacitance of about 0.2 pF, was obtained: and the effective dielectric constant ε_{eff} of 16 was obtained by comparison with a capacitance of 0.044 pF in free space, calculated with conformal mapping techniques. For higher operation, the travelling wave electrodes as shown in Fig. 8 are suitable. In Fig. 8, the characteristic impedance is designed to match 50 ohms at both ends of the electrodes assuming that $\varepsilon_{eff}=16$. Fig. 9 shows a frequency response of the return loss in the travelling wave electrodes. From Fig. 9, an efficient impedance matching is realized in the frequency range 45 MHz to 26.5 GHz. This proves that it is reasonable to assume the dielectric constant as 16. The effective refractive indices were obtained as $N_0=2.5$ in optical waves and $N_m=\varepsilon_{eff}^{\frac{1}{2}}=4$ in microwaves. The band width of the TIR optical switch with the travelling wave electrodes as shown in Fig. 8 is roughly estimated from the equation,

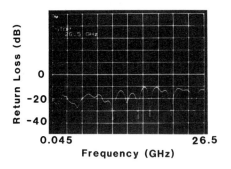

Fig. 7. Modulated light wave form from TIR switch (type II).

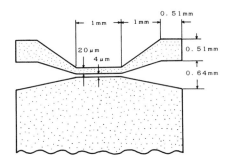

Fig. 8. Travelling wave electrodes.

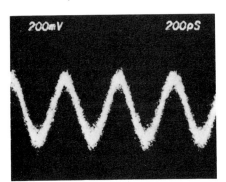

Fig. 9. The frequency response of the return loss in travelling wave electrodes.

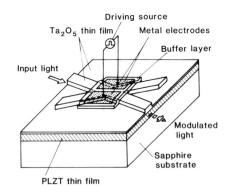

Fig. 10. Construction of IWOS.

$$\Delta F = 1.9c/(\pi \cdot |N_m - N_o| \cdot L) \tag{3}$$

where the microwave loss is neglected, c is the light velocity in free space, and L is the length of the electrodes. The calculated bandwidth with L=1 mm is over 100 GHz. In the above estimation, the velocity variation of the guided light in the intersection was not considered.

4. INTERSECTING WAVEGUIDE-INTERFERENCE-TYPE OPTICAL SWITCH

The TIR-type optical switches are easy to fabricate but their operation is not stable because of their multimode operation. To obtain stable operation of the switch, single- or a few-mode operation is required. In this operation mode, total internal reflection hardly occurs, but light waves of the guided modes in the intersecting waveguide optical switch (IWOS) interfere with each other. The output lights become more sensitive to the intersection structure and more precise control of the structure is required. Higashino et al(1987) have developed the IWOS and carried out an optical signal transmission experiment using IWOS as an external modulator.

4.1 STRUCTURES AND DESIGN

The configuration of the IWOS is shown in Fig. 10(Higashino et al 1987). In this switch, variable thickness cladding waveguides are adopted. The ridge waveguide has a disadvantage in causing unnegligible scattering loss at irregular side walls, because the PLZT thin film has a high refractive index of 2.6. On the other hand, the strip-loaded waveguide has an advantage in the control of propagation constants. However, the surface of the waveguide except for the strip region is left as a bare surface of PLZT film. In Fig. 10, the PLZT film is covered by a Ta_2O_5 cladding layer, therefore, the surface scattering of the guided light becomes much smaller. The Ta_2O_5 cladding is also covered by a buffer layer in the intersecting region. The buffer layer has a lower refractive index than that of the cladding.

The intersecting region can be thought of as strongly coupled dual channel waveguides whose coupling strength and propagation constants are changed by a driving electric field application. The switching characteristics are simulated by the anomalous numerical calculations using the effective refractive method(Marcatili 1969 and Furuta 1974). However, this is not suitable for the design because of its long calculation time. We have developed an approximate designing method for IWOS. The details of this will appear in another paper(Higashino et al submitted for publication). In this section, we introduce a summary of the method as follows.

For simplification of the analysis, we introduce a model of a coupled waveguide structure as shown in Fig. 11, where W_0 is the width of the input and the output waveguides and W_1 is the minimum width of the intersection. The intersecting angle is ϕ and n_1, n_2 and n_g are effective refractive indices in the guiding region, outside the region and the gap between, respectively. The guided light field is represented by fields of even and odd modes in the infinitesimal dz region at a distance z. The propagation constant $\beta(z)$ varies along the z direction. When a voltage V is applied to the electrodes, the effective refractive index of the gap region is changed. Therefore, the difference in propagation constants $\Delta\beta(z)$ between the even mode and the odd ones is changed and is approximated by

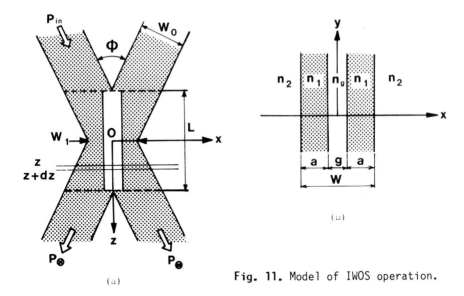

Fig. 11. Model of IWOS operation.

$$\Delta\beta(z)=(\beta_{e0}(z)-\beta_{o0}(z))\cdot\exp(A(\Delta n_g)\cdot\Delta n_g), \qquad (4)$$

where $\Delta n_g=n_g(V)-n_1$ and the subscript "0" denotes the case that $n_g=n_1$. The subscripts "e" and "o" denote even and odd mode, respectively. The value of $A(\Delta n_g)$ is changed by V and in the coordinate of z. The total phase shift $\Omega(\Delta n_g)$ between the even and the odd modes is approximated as

$$\Omega(\Delta n_g)=\int_{-L/2}^{L/2}\Delta\beta(z)dz=2\int_{0}^{L/2}\Delta\beta_0(z)\cdot\exp(A(\Delta n_g)\cdot\Delta n_g)dz$$

$$=L\cdot\Delta\beta|_{W=1.5W_1}. \qquad (5)$$

If the incident light wave of unit optical power P_{in} is travelling into the intersection at $z=-L/2$, through power $P_\ominus$ and cross power $P_\otimes$ at the end of the intersection at $z=L/2$ are obtained as follows:

$$P_\ominus=\sin^2(\Omega(\Delta n_g)/2), \qquad (6)$$

$$P_\otimes=\cos^2(\Omega(\Delta n_g)/2). \qquad (7)$$

To obtain the required characteristics, several approximate calculations and designs are carried out. After determination of the proper set of effective refractive indices, the determination of structure constants for the waveguides is left to the last design where the effective refractive indices are obtained. In Table 2, the design constants are listed.

Fig. 12 shows the variation of $\Delta\beta$ versus Δn_g at 0.83 µm. The solid line shows the numerical solution and the broken line does the approximate one. The output power variation is shown in Fig. 13. From these results, the propriety of the approximate designing method is proved.

Table 2. Design parameters of the PLZT-IWOS

Layer	Refractive index	Thickness (µm)
Substrate	1.77 (n_s)	—
PLZT thin film	2.6 (n_f)	0.35 (h)
Cladding	2.09 (n_c)	0.01 (t_1)
Loaded strip	2.09 (n_c)	0.03 (t_2)
Buffer	1.87 (n_b)	0.18 (t_b)

Waveguide;	W_0=10 µm	ϕ=1°
Electrode;	g=4 µm	L=2.1 mm

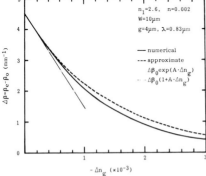
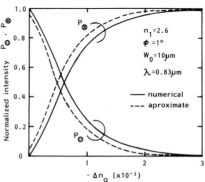

Fig. 12. $\Delta\beta$ vs. Δn_g characteristics.   **Fig. 13.** $P_\ominus$, $P_\otimes$ vs. Δn_g.

4.2 FABRICATION AND SWITCHING CHARACTERISTICS

The designed IWOS in Table 2 was fabricated using the techniques as mentioned in section 3.1. The measurements were carried out using a LD of 0.83 µm wavelength. The insertion loss of the intersection with a length of 1.2 mm was much less than 1 dB. The minimum branching ratio of the switch was measured as about −20 dB. The IWOS was assembled into a module with a polarization maintaining fiber (PMF) of 5 µm core diameter for input and a 10 µm core optical fiber for output. These fibers were coupled to the input and the output waveguides through an GRIN lens and a GaP prism. The input and output waveguides were in multimode, therefore, coupling was carried out to obtain the maximum extinction ratio. The fibers and lenses were fixed with a low shrinking adhesive resin. The measured total insertion loss of the module was 36 dB which was mainly due to the input/output coupling losses. The coupling loss can be reduced by fabricating a large core waveguide which has a graded refractive index structure and is made by a multi-target sputtering system(Adachi et al 1985). The electrodes of the IWOS were connected with bonding wires to the terminated strip line.

Fig. 14 shows the switching characteristics of the module at 0.83 µm. The origin of the measured curve (solid line) seems to shift as 5 V. This can be thought of as due to residual polarization of the PLZT film. The dotted line shows the numerical solution by putting $R=1\times10^{-17}$ (m/V)^2. The theoretical curve is well fitted to the measured one except for under −13

V. From this result, it can be seen that the strength of the electric field in the PLZT film was less than half of the apparent strength in the electrode gap. The extinction ratio was 12.5 dB for 18 V_{p-p} application with -4 V-dc bias.

The 3 dB cut-off frequency of the return loss was 3.4 GHz, which was determined mainly by the series LC resonance of the bonding wire inductance and the electrode capacitance. Fig. 15 shows the frequency characteristics of the modulator which include those of a driving amplifier, the detector and·the preamplifier. The overall frequency characteristics of the modulation depth and the group delay were as follows: less than 1.2 dB deviation up to 200 MHz and less than 1 nsec deviation up to 1 GHz. In this structure, the interaction length between the optical wave and the electric field was about 1.4 mm, considering evanescent coupling in the outside region of the intersection. If travelling wave electrodes are used in this IWOS, a modulation bandwidth of over 80 GHz can be expected, from equation (3).

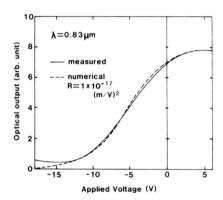
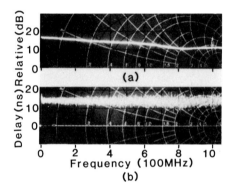

Fig. 14. Optical output of IWOS vs. applied voltage.   **Fig. 15.** Frequency response of IWOS.

4.3 OPTICAL TRANSMISSION EXPERIMENT

Because the PLZT thin films have a strong quadratic E-O effect, the PLZT-IWOS is suitable for optical digital transmission. To examine the PLZT-IWOS, optical digital signal transmission experiments were carried out at a constant temperature for more than 1000 hours(Higashino et al 1987). The IWOS modulator was used as a modulator and the output fiber was connected to a 1 km multimode optical fiber (GI-50/125). The driving signal rate was 200 Mbit/sec (return-to-zero) and the driving power level was 24 dBm with a dc bias. The transmitted signals were detected with a APD. Fig. 16 shows the driving signal wave form and the received optical signal wave form after 1 km transmission. The deviation of the received average optical power was less than 0.4 dB and the received signal level was hardly changed during the experiments.

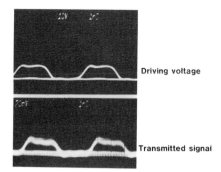

Driving voltage

Transmitted signal

Fig. 16. Driving and transmitted wave form.

5. CONCLUSION

The fabrication process of PLZT thin films and their properties are described, and two kinds of TIR-type optical switches are introduced. These switches exhibited a low driving voltage of 4.7 V and a high speed operation of 2 GHz. Interference-type intersecting waveguide optical switches were more stable in operation than the TIR-type switches.

The layered structure of PLZT thin film/sapphire is found to be useful for optical waveguide switches. Further study will bring success in a realization of high-speed optical signal processors, using the PLZT switches for an optical communication system.

ACKNOWLEDGMENTS

The authors would like to thank Dr. T. Kawaguchi, Dr. T. Makino and Dr. O. Yamazaki for their collaborations and useful discussions and Dr. T. Nitta for his continuous encouragement. They would also like to thank Dr. H. Nakata, Dr. M.Tanabe, Dr. K. Utsumi, Dr. K. Fujito and Dr. T. Ichida for their collaboration on the transmission experiments, and Dr. S. Ishizuka for his advice on the LD module fabrication.

REFERENCES

SECTION 1
Adachi H, Kawaguchi T, Setsune K, Ohji K and Wasa K 1983a: Appl. Phys. Lett. **42** pp 867-8
Haertling G H and Land C E 1971: J. Am. Ceram. Soc. **54** pp 1-11
Higashino H, Kawaguchi T, Adachi H, Makino T and Yamazaki O 1985: Japan. J. Appl. Phys. **24-2** pp 284-6
Higashino H, Adachi H, Nakata H, Tanabe M, Utsumi K, Fujito K and Yamazaki O 1987: Technical Digest of Optical Fiber Communication Conf./6th Int. Conf. on Integrated Optics and Optical Fiber Communication **TUQ19**
Ishida M, Matsunami H and Tanaka T 1977: Appl. Phys. Lett. **31** pp 433-4
Ishida M, Tsuji S, Kimura K, Matsunami H and Tanaka T 1978: J. Cryst. Growth **45** pp 393-8
Okuyama M, Usuki T, Hamakawa Y and Nakagawa T 1980: Appl. Phys. **21** pp 339-43
Tsai C S, Kim B and El-Akkari F R 1978: IEE J. Quantum Electron. **QE-14** pp 513-7
Wasa K, Yamazaki O, Adachi H, Kawaguchi T and Setsune K 1984: IEEE J. Lightwave Tech. having **LT-2** pp 710-714

SECTION 2

Adachi H, Kawaguchi T, Setsune K, Ohji K and Wasa K 1983a: Appl. Phys. Lett. **42** pp 867-8

Adachi H, Kawaguchi T, Kitabatake M and Wasa K 1983b: Japan J. Appl. Phys. **Suppl. 22-2** pp 11-3

Adachi H, Mitsuyu T, Yamazaki O and Wasa K 1986: J. Appl. Phys. **60** pp 736-41

Kawaguchi T, Adachi H, Setsune K, Yamazaki O and Wasa K 1984: Appl. Optics **23** pp 2187-91

Nakagawa T, Yamaguchi J, Usuki T, Matsui Y, Okuyama M and Hamakawa Y 1979: Japan J. Appl. Phys., **18** pp 897-902

Okuyama M, Usuki T, Hamakawa Y and Nakagawa T 1980: Appl. Phys. **21** pp 339-43

Wasa K, Yamazaki O, Adachi H, Kawaguchi T and Setsune K 1984: IEEE J. Lightwave Tech. **LT-2** pp 710-714

SECTION 3

Hammer J M 1975: Integrated Optics ed Tamir T (New York Heidelberg Berlin: Springer-Verlag) pp 175-189

Higashino H, Kawaguchi T, Adachi H, Makino T and Yamazaki O 1985: Japan. J. Appl. Phys. **24-2** pp 284-6

Tsai C S, Kim B and El-Akkari F R 1978: IEE J. Quantum Electron. **QE-14** pp 513-7

Wasa K, Yamazaki O, Adachi H, Kawaguchi T and Setsune K 1984: IEEE J. Lightwave Tech. **LT-2** pp 710-714

SECTION 4

Adachi H, Mitsuyu T, Yamazaki O and Wasa K 1985: Japan J. Appl. Phys. **Suppl.24-3** pp 13-6

Furuta H, Node H and Ihaya A 1974: Appl. Optics **13** pp 322-6

Higashino H, Adachi H, Nakata H, Tanabe M, Utsumi K, Fujito K and Yamazaki O 1987: Technical Digest of Optical Fiber Communication Conf./6th Int. Conf. on Integrated Optics and Optical Fiber Communication **TUQ19**

Higashino H, Adachi H, Nakata H, Tanabe M, Utsumi K, Fujito K and Yamazaki O submitting to IEEE Lightwave Tech.

Marcatili E A J 1969: Bell Syst. Tech. J., **48** pp 2071-102

Inst. Phys. Conf. Ser. No 103: Part 1
Paper presented at Int. Conf. Materials for Non-linear and Electro-optics, Cambridge, 1989

MOCVD growth of inorganic thin films

C J BRIERLEY & C TRUNDLE

Plessey Research & Technology, Caswell, Towcester, Northants, NN12 8EQ

ABSTRACT

Investigation into the growth of inorganic thin film materials has
increased rapidly in recent years for applications as diverse as
tribological coatings for tool steels to optical coatings for semiconductor lasers. Such coatings are generally composed of either a
metal oxide, nitride or carbide, but recently, interest has
focussed on thin films of more complex oxide materials such as
those used in high temperature superconductor and non linear
electro-optic applications.

Traditional processes such as sputtering have been used and proven,
but many important materials have been found to be difficult to
deposit as coherent crystalline films using such physical vapour
deposition techniques. An alternative route to depositing good
quality crystalline oxide materials is Metal Organic Chemical
Vapour Deposition (MOCVD), a technique that has been used with
considerable success in the deposition of compound semiconductors,
but with little reported on its use in the deposition of complex
oxides.

This paper describes how a range of newly synthesised
organo-metallic precursors have been used to develop an MOCVD
process capable of growing thin films of transparent ferroelectric
oxides such as $PbTiO_3$ and how this process might be used to produce
more useful non-linear electro-optic oxides.

1. INTRODUCTION

It has become increasingly desirable to deposit inorganic electro-optic
materials directly in thin film forms for many optical applications. In
particular, the trend towards microlithographic techniques and monolythic
processing has enabled many optical functions to be integrated into the
surface of a single slice of material. Such slices are conventionally
produced by sawing, thinning and polishing bulk single crystal materials.
Such processes are both costly and wasteful of material. In addition,
the requirement is increasingly to integrate the optical functions with
electronic functions. This would most easily be achieved if the
optically active material is deposited directly onto the substrate
containing the signal processing functions (ie, silicon).

LiNbO$_3$ is an example of an important electro-optic material used in the production of complex optical switching devices, which integrate many functions into one slice of material. The benefits of being able to grow such a material onto a silicon substrate, in the thickness and form required are very great. Many other similar oxide materials which possess greater electro-optic activity than LiNbO$_3$ are known, but are not readily available as bulk single crystals. Examples include (BaSr)$_1$ Nb$_2$O$_6$ and K(NbTa)O$_3$ among others. Many of these may lend themselves to growth as thin film crystalline films. Another such material in this case having an exceptional quadratic electro-optic coefficient is lead lanthanum zirconate titanate (PLZT). This has never been grown as single crystal material, but only as polycrystalline ceramic. Recent work has demon- strated that single crystals (or orientated polycrystalline) material can be grown on suitable epitaxial substrates by sputtering (1 & 2), and indeed the general process of sputtering has become one of the most successful thin film deposition techniques for materials which cannot be congruently evaporated (3).

The use of sputtering in growing oxides with particular crystalline properties, however, has not always proved successful, and indeed, is far from easy with PLZT. For this reason, many other thin film deposition techniques have been investigated and the recent growth of interest in the deposition of high temperature superconductors such as yttrium barium copper oxide has given rise to research into the thin film deposition of complex oxides by a very wide range of thin film techniques. Reference to the proceedings of a Materials Research Society Meeting in 1987 (4) reveals that no less than 24 papers were devoted to different processes that had been succesfully used in the thin film deposition of high Tc superconductors. These included rf, magnetron, dc and ion beam sputtering, sol-gel deposition, laser and e-beam evaporation, MBE, spray pyrolsis and plasma spraying. This list is remarkable in its diversity but significant for the lack of any reference to chemical vapour deposition techniques.

Chemical vapour deposition processes are widely used in the deposition of simple metal oxides, nitrides and carbides in applications varying from the tool steel industry (5) to the semiconductor industry (6). In recent years, the Metal-Organic Chemical Vapour Deposition (MOCVD) of compound semiconductors has become one of the major processing technologies (7) for controlled deposition of epitaxial films. It is therefore surprising that only very recently, some reports of the MOCVD of superconductors and other ferroelectric materials have been published (8, 9). The delay can largely be explained by the lack of commercially available volatile metal organic precursors for many of the target metals required and the relative toxicity associated with some classes of metal-organics notably the metal alkyls. Metal oxides have, in fact, been more commonly deposited using the metal alkoxides rather than metal alkyls as precursors. C C Wang et al (10), for instance, have described the deposition of Ta$_2$O$_5$, and TiO$_2$ from their respective metal ethoxide derivatives. However, because of the lack of alkoxide precursors for Nb and Pb, they resorted to the use of NbOCl$_3$ and PbO to transport these metals, neither of which are very convenient to use. These difficulties highlight why very little has been reported in the past on the deposition of complex oxides by CVD at all. M Kojima et al (11) was the first to describe the growth of PbTiO$_3$ in this case from TiCl$_4$, PbCl$_2$ and H$_2$O, but

the resulting films were of poor crystalline quality with considerable
incorporation of chlorine.

This paper describes the progress made in developing MOCVD processes for
PbTiO$_3$. The key to much of this work has been the ability to design and
synthesise, in-house, volatile metal organic precursors which are not
available commercially.

2.   THE DEVELOPMENT OF METAL ORGANIC PRECURSORS

The deposition of III-V and II-VI semiconductor compounds by MOCVD has
necessitated the development of precursors with the rigorous exlusion of
oxygen from their structures and in the carrier gas during growth. The
metal alkyls and hydrides have been almost exclusively used. These
materials have the advantage of being reasonably volatile at room
temperature or below, but are also very toxic and often pyrophoric.

Metal oxide deposition clearly does not require oxygen free precursors
and as a result, many workers have opted to use the metal alkoxides or
acetyl acetonates instead (12-14). These tend to be less volatile than
their alkyl equivalents, but are easier to handle with a much lower
toxicity. K J Sladek and W Gibert (14) reported the deposition of Al$_2$O$_3$,
Nb$_2$O$_5$, Sb$_2$O$_3$, TiO$_2$ and ZrO$_2$ from their respective ethoxides and butoxides
by low temperature hydrolysis, but found the conditions necessary for
good quality thin film deposition were difficult to control. This is not
an uncommon observation with atmospheric deposition systems where
interfering homogenous reactions resulting in powder formation are
commonly observed (14). We have found that deposition systems using
alkoxide precursors perform much better at low pressures and that reliable
thin film pyrolysis readily occurs at 350-400^{0}C on to a substrate without
any competing homogenous reactions.

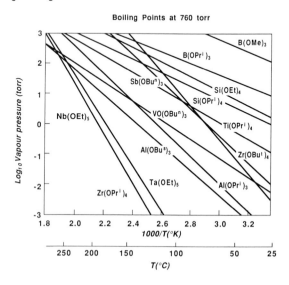

Figure 1   Volatilities of some commercially available metal alkoxides
(Taken from Ref. 32)

Figure 1 shows the volatilities of some commercially available oxides (taken from Ref 12). It will be noted that most of them with the exception of the boron alkoxides and silicon alkoxides require heating to achieve a vapour pressure of 0.1torr at which efficient low pressure transport of the precursor will occur. The volatility of alkoxides is dependent on a number of factors which are clearly described in Bradley and Mehrotra's excellent volume (13). Basically, volatility is related to the molecular weight of the compound. However, since alkoxides have a strong tendency to oligomerise through the co-ordination of the metal atom of one molecule with an oxygen atom on adjacent molecules, they are rarely found as monomers. The $Al(OPr^i)_3$ forms a trimer of the type shown in Figure 2a, in solid, liquid or vapour (identified by X-ray and NMR) but reverts to the tetramer (Figure 2b) on ageing. The degree of oligomerisation that can take place is reduced by steric factors such as the number and complexity of the ligands attached to the metal. It will be noted from Figure 1 that no alkoxides with a valency of less than 3 is included in this figure. This is quite simply because divalent and monovalent metals tend to oligomerise to such an extend that they become largely involatile. Large branched alkoxides can, however, reduce this effect by 'wrapping the metal up'. Thus, contrary to intuition, large alkoxide groups can considerably increase volatility. An example of this is found in the Pb(II) alkoxides. $Pb(OPr^i)_2$ and $Pb(OBu^n)_2$ can be synthesised, but neither can be distilled or sublimed at low pressure. In contrast, the heavily branched tertiary butoxide $Pb(OBu^t)_2$, which may be synthesised by the route described by B Cetinkaya et al (14), can be readily sublimed at 130^0C at 0.1torr. Such complex alkoxides are generally not available commercially and must be synthesised. We have synthesised this and related metal alkoxides using the route described in reference 14.

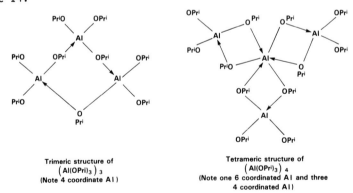

Trimeric structure of
$(Al(OPri)_3)_3$
(Note 4 coordinate Al)

Tetrameric structure of
$(Al(OPri)_3)_4$
(Note one 6 coordinated Al and three 4 coordinated Al)

Figure 2   Oligomeric structures of $Al(OPr^i)_3$

An alternative adduct that has been used for the synthesis of volatile precursors involves a β-di-ketonate liquid. The metal acetylacetonates are generally more stable than the alkoxides, being pyrolytically decomposed at temperatures in excess of 450^0C while not being hydrolysed by water at normal temperatures. J Dismukes et al (15) report the use of β-di-ketonates in the spray pyrolysis of Y_2O_3. Figure 3 shows a typical reaction route for the synthesis of these compounds and their structure. By modifying the substituents at R and R' the volatility can be increased, particularly if fluorinated groups are substituted. Such compounds, by increasing the co-ordination number achieved by the metal atom, are less prone than alkoxides to high degrees of oligomerisation, and are often therefore more volatile. These organo-metallic precursors are now becoming increasingly available commercially, particularly for the metals involved in the manufacture of high temperature superconductors.

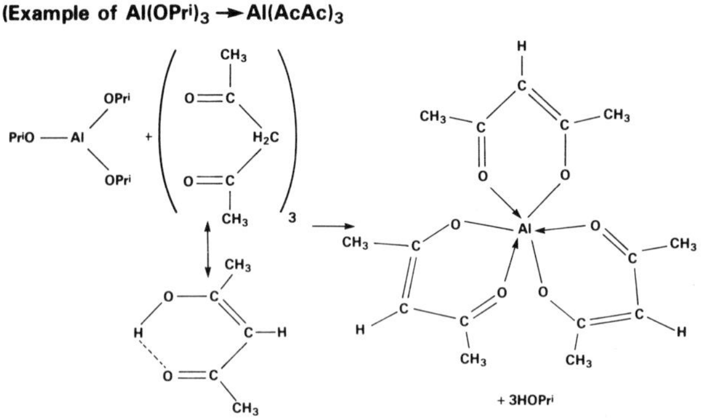

(Example of Al(OPri)₃ → Al(AcAc)₃

Figure 3 Reaction scheme for the synthesis of metal acetyl acetonates

The work described below in the deposition of PbTiO$_3$ started with two alkoxides, titanium isopropoxide and lead tertiary butoxide, but two acetyl acetonates are more commonly used today, namely lead bis dimethyl heptafluoroacetyl acetonate and titanium di-isospropoxide di-acetyl acetonate. The former is more volatile than the lead tertiary butoxide and melts at temperatures just above ambient. This means it is easier to transport as a vapour from a warm bubbler than the lead butoxide which can only be sublimed giving less consistent evaporation rates. The titanium mixed alkoxide/acetyl acetonate is, by comparison, less volatile than the simple isopropoxide, but has the advantage of being more resistant to hydrolysis and therefore easier to handle in the atmosphere.

From the foregoing, it can be seen that relatively volatile metal organic compounds of a substantially non-toxic nature can readily be prepared for selected metal oxides and in this way, the potential for the MOCVD of complex electro-optic oxides is clearly available.

3. DEPOSITION APPARATUS

The deposition apparatus used in this work is shown schematically in Figure 4. The system operates at pressures below atmospheric for the reasons outlined in the previous section. In order to achieve a sufficient and well controlled vapour pressure of the precursors, the metal-organics are held at a precise temperature in stainless steel or glass bubblers heated by means of a recirculated oil jacket. The gas lines are also heated to avoid condensation in the pipework. Deposition is achieved in a silica tube with an inductively heated encapsulated

graphite susceptor as shown in Figure 4. This arrangement allows
temperatures of up to 900⁰C to be achieved.

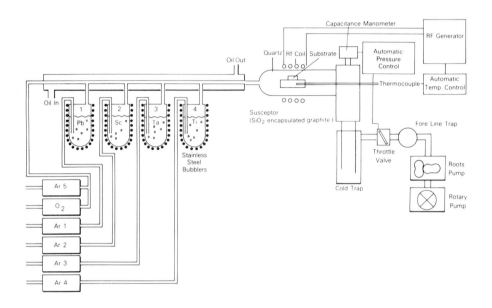

Figure 4   MOCVD Deposition Apparatus

4.   DEPOSITION OF PbTiO₃

Under typical conditions of 1Torr pressure, total flow rates of 100SSCM
and substrate temperatures of 400-700⁰C films of either PbO, TiO₂ or
mixtures of the two may be deposited in the presence of an inert carrier
gas such as argon. It is observed that under these conditions, the
oxides are formed more readily from the alkoxides than the acetyl
acetonates, which tend to form films heavily contaminated with carbon

At deposition temperatures of 400-500⁰C, the mixed oxides do not
generally form well defined crystalline phases and may often be
amorphous. At such deposition temperatures, it has not been possible to
directly deposit the perovskite phase. Instead, a layer of the desired
thickness of an intimately mixed combination of metal oxides of the
correct composition are deposited. Subsequent annealing in oxygen
produces the desired crystalline phase of PbTiO₃.

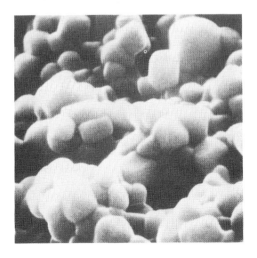

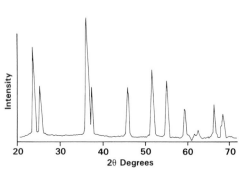

Figure 5 PbTiO$_3$ gain size 0.5 - 1μm

Figure 6 X-ray diffraction traces of MOCVD grown PbTiO$_3$

This process is best illustrated as follows: a 1:1 ratio of PbO and TiO$_2$ is deposited on to sapphire at 450^{0}C from Pb (OBu^t)$_2$ and TiOPr^i)$_4$ up to a thickness of 5μm. Annealing this at 800-900^{0}c in air produces a film of polycrystalline PbTiO$_3$ with a grain size of approximately 0.5-1.0μm (Figure 5). The X-ray diffraction pattern of this material is shown in Figure 6 and clearly shows the perovskite phase of PbTiO$_3$. The grain size can be varied by changing the deposition conditions and annealing temperature. An example of this is shown in Figure 7, which shows some SEM micrographs of PbTiO$_3$ with a grain size of up to 25μm after annealing at 1100^{0}C in an overpressure of PbO.

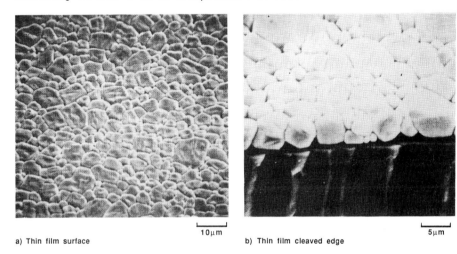

a) Thin film surface

b) Thin film cleaved edge

Figure 7 Example of MOCVD grown PbTiO$_3$ annealed at 1100°C

Ideally, however, one would wish to deposit the perovskite phase by a single step deposition process, this may be accomplished by adding a hydroxyl source into the gas stream such as water or alcohol. This does not give good results with alkoxides which tend to break down prematurely, but is much more successful with the acetyl acetonates whose breakdown to oxides at 400^0C is considerably enhanced by the addition of water as the volatile β-diketone group is able to leave cleanly as illustrated below for $Pb(Acac)_2$

$$Pb\begin{bmatrix} & & CH_3 \\ & & / \\ O & - & C \\ / & & \backslash\backslash \\ & & CH \\ & & / \\ O & = & C \\ & & \backslash \\ & & CH_3 \end{bmatrix}_2 + H_2O \longrightarrow PbO + 2\begin{bmatrix} & & CH_3 \\ & & / \\ O & = & C \\ & & \backslash \\ & & CH_2 \\ & & / \\ O & = & C \\ & & \backslash \\ & & CH_3 \end{bmatrix}$$

However, the increased oxygen potential of the atmosphere also allows the PbO and TiO_2 to react in situ to form the crystalline form $PbTiO_3$. An example of perovskite grown in this way at 700^0C is shown in Figure 8. This film, grown onto sapphire is approximately $5\mu m$ thick and highly transparent. The sub-micron grain size does not scatter light. An interdigitated metal pattern has been applied to the surface to allow electrical measurements to be made.

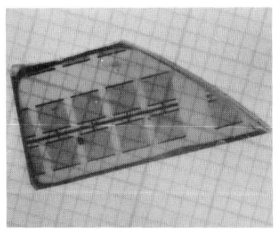

Micrograph of same material
(note uniformity of grain size)

As grown $PbTiO_3$ with metal
transducers applied

Figure 8   Example of optical quality $PbTiO_3$ as grown by MOCVD

5.   TOWARDS MOCVD OF ELECTRO-OPTIC THIN FILMS

While MOCVD has successfully been applied to the growth of thin films of polycrystalline $PbTiO_2$, second order electro-optic materials such as

LiNbO$_3$ must be deposited in a non centro-symmetric fashion with respect to their optical axis, so that the desired change in birefringence occurs uniformly within the film. In practice, this means that inorganic thin film materials must be grown as a single crystal, non centro-symmetric phase. In certain situations, a strongly orientated polycrystalline thin film may suffice provided the grain boundaries do not significantly interfere with the optical performance. In general, therefore, an epitaxial relationship between substrate and thin film is required to encourage single crystal growth (16), and therefore a single crystal substrate material having lattice parameters close to those of the deposited layer (in the plane of thin film crystal) must be chosen.

In vapour deposition processes in general, a lattice mismatch of 5% or sometimes greater can be tolerated. As a further requirement of successful epitaxy, the energy requirements of crystal growth must be considered. For crystal growth to be successful, it is essential for the depositing species to have sufficient energy to achieve mobility upon the substrate surface, allowing migration of the species to a growing crystal step. As a rule of thumb, a thermal energy of more than 2/3 the absolute melting point is necessary to achieve this. In the case of LiNbO$_3$ this implies a substrate temperature in excess of 750^{0}C, and because many oxides are even more refractory, these deposition temperatures are often undesirable or impossible to achieve. Various surface stimulation processes, however, are available, which can considerably increase the energy of surface absorbed species while maintaining a relatively cool substrate. A typical RF or DC plasma achieves this under normal operating conditions because either electron or ion bombardment of the growing surface will occur from the gas phase depending on the electrode biasing arrangements. This is one reason why sputtering of oxides has proved so popular, because relatively high temperature phases of oxides can be deposited onto substrates at temperatures below that required in other forms of CVD. However, other more direct means of surface stimulation are now gaining in popularity, including ion beams, atomic beam and optical beams. Each of these have been used with success to enhance crystal growth in CVD systems (17-19). Both ionic and atomic beams are similar in concept in so far as the accelerated particle hits the surface of the substrate and exchanges energy through the 'billiard ball' collision process. Weakly bound species are resputtered from the surface or are able to migrate on the surface to a more favourable binding site, such as a growing epitaxial step. Under optimum conditions only those molecules in a crystal lattice are able to survive. Provided a good epitaxial relationship exists between the substrate and a number of growing nuclei, then these will ultimately coalesce to form a perfectly ordered crystalline surface.

Optical beams operate in an analogous fashion. Photons are absorbed by the surface adsorbed species ultimately raising their kinetic energy. UV light is generally used as this is readily absorbed by most materials and a single photon absorption will often excite the surface absorbed species to a higher electronic energy level. The adsorbed species are then free to migrate to a energetically more favourable binding site or desorb altogether, in a similar fashion to that described above.

The advent of powerful UV lasers such as the excimer, means that the stimulating photon energy (wavelength) may be chosen from a number of possible emissions between 193 and 400nm. The use of photons also means that surface stimulation may be carried out in a very wide pressure window from ultra high vacuum to atmospheric pressure or above. Atomic and ion beams by their nature must be operated in a fairly limited pressure window. Growth techniques such as molecular beam epitaxy (MBE) have demonstrated that many materials can be grown more successfully and with greater crystalline perfection at ultra high vacuums, and it follows that epitaxial deposition by CVD should benefit from as low pressure growth as possible. For this reason, photo enhanced MBE is a developing technology (20).

It is therefore clear that while some work needs to be done before MOCVD of electro-optic films becomes a reality, the objective of epitaxy at modest temperatures is achievable; the main requirement being to achieve epitaxy at temperatures that are compatable with the chosen substrates. This temperature will obviously vary with the application, but certainly for many integrated electro-optic devices a temperature consistent with signal processing chips would seem desirable, ie, not more than $500-600^0$ C. It is likely that most oxide deposition processes will require a degree of surface stimulation before this can be achieved.

6.   CONCLUSIONS

The development of MOCVD for the deposition of complex oxides has been delayed by the lack of available volatile metal organic precursors. This problem is now being addressed for the metals found in current high temperature superconductors. We have developed a range of volatile precursors suitable for $PbTiO_3$ deposition and some other ferroelectric materials. MOCVD is found to occur best at low pressures onto substrates heated above 450^0 C. Reactions between oxides are improved in the presence of hydroxyl groups such as water vapour or alcohol. Polycryst-alline layers of $PbTiO_3$ of up to $10\mu m$ thick have been deposited onto substrates such as sapphire. The grain size of a crystal phase may be varied by the exact deposition conditions and various post deposition annealing steps.

For this process to be compatible with the growth of non-linear electro-optic materials such as $LiNbO_3$ it is desirable to deposit single crystal epitaxial films in order to achieve non-centrosymetry of the molecules and to avoid optical scattering at grain boundaries. This implies the use of substrates with a good epitaxial relationship with the desired crystal orientation and somewhat more energetic deposition conditions than those currently used. Various techniques likely to achieve this are discussed which do not involve excessive substrate heating, but rather stimulate the growing film to encourage epitaxy to occur.

ACKNOWLEDGEMENTS

This work has been supported and sponsored by the procurement executive, Ministry of Defence (Royal Signals and Radar Establishment)

REFERENCES

1. A Okada, 'Some electrical and optical properties of ferroelectric PZT thin films', Jour Appl Phys, Vol 48, No 7, (1977), pp2905-2908.

2. H Adachi, T Kawaguchi, IK Selsane, K Ohiji & K Wasa, 'Electro-optic effects of (Pb, La)(Zr, Ti)O_3 thin films preapred by rf planar magnetron sputtering', App Phys Lett, 42, 10 (1983), pp867-868.

3. B Chapman, 'Glow discharge processes', Published John Wiley & Sons, 1980 Chapter 6, pp177-284.

4. M B Brodsky, R C Dynes, K Kitazawa & H L Tuller, 'High temperature superconductors', MRS Proceedings, Vol 99, 1988.

5. M T Laughier, 'Adhesion of TiC and TiN coatings prepared by chemical vapour deposition on Wc-Co-based cemented carbides', Jour Mat Sci, 21, (1986), pp2269-2272.

6. J A Amuk, G L Shnable & J L Vossen, 'Depositon technology for dielectric films in semiconductor devices', J Vac Sci Technol, 14, (1977), pp1053-1153.

7. M J Ludowise, 'Metal organic chemical vapour deposition of III-V semiconductors', J Appl Phys, 58, No 8, (1985), ppR31-R55.

8. S Oda, H Zama, T Oshtsuka, Z Sugiyama & T Hattori, 'Epitaxial growth of YBaCuO films on sapphire at 500^0C by MOCVD' Jap Jour App Phys, Vol 28, 3, (1989), ppL427-L429.

9. S L Swartz, D A Seifant & G T Neol, 'Characterisation of MOCVD PbTiO$_3$ thin films', Ferroelectrics (1989).

10. C C Wang, K H Azninger & M T Duffy, 'Vapour deposition and characterisation of metal oxide thin films for electronic applications', RCA Review Dec 1970, pp728-741.

11. M Kojima, M Okuyama, T Takagawa & Y Hamakawi, 'Chemical vapour deposition of PbTiO$_3$ thin film', Jap Jour Appl Phys, Vol 22, (1983), Supplement 22-2, pp14-7.

12. K J Sladek & W Wayne Gilbert, 'Low temperature metal oxide deposition by alkoxide hydrolysis', proc III Int Chem Vapour Deposition Conference, (London), (1978).

13. D Bradley, R C Mehrotra & D P Gauer, 'Metal alkoxides', (1980), p2088.

14. B Cetinkaya, I Gumrukcu et al, J Am Chem Soc, 102:6, (1980), p2088.

15. J P Dismukes, J Kane et al, 'Chemical vapour deposition of cathodoluminescent phosphor layers', Proc IV Int Chem Vapour Deposition Conference Eds G F Wakefield, J M Blocher (1983).

16. R C Pond, ' Crystallographic analysis of domain formation in epitaxial films', Jour Cryst Growth, 79 (1986), pp946-950.

17. J M E Harper, 'Ion beam techniques in thin film deposition', Solid State Technology, April (1987), pp129-134.

18. H Kuwano & K Nagai, 'Friction reducing coatings by dual fast atom beam technique', J Vac Sci Technol A, Vol 4, 6, (1986), pp2993-2996.

19. T kawai, T Choda & S Kawai, 'Laser induced formation of thin TiO_2 films from Ti_2Cl_4 and oxygen onto a silicon surface', MRS Symp proc, Vol 75, (1975), pp289-295.

20. A Doi, Y Aoyagi & S Namba, 'Growthy of GaAs by switched laser metal- organic vapour phase epitaxy', App Phys Lett, Vol 48, 26, (1986), pp1787-1789.

Periodically domain-inverted waveguides in lithium niobate for second harmonic generation: influence of the shape of the domain boundary on the conversion efficiency

Gunnar Arvidsson and Bozena Jaskorzynska

Institute of Optical Research, S-100 44 Stockholm, Sweden

ABSTRACT: It has recently been reported that periodically domain-inverted waveguides have been fabricated in $LiNbO_3$ and used for quasi-phase-matched second harmonic generation. This opens up possibilities for a much broader use of $LiNbO_3$ waveguides for second order nonlinear processes. We review experimental results, and present calculations on the conversion efficiency. The applied theoretical model is based on coupled-mode-theory and accounts for different shapes and periods of the inverted domains. It is concluded that with optimized waveguide structures it should be possible to realize laser diode pumped frequency doublers with output powers in the mW range.

1. INTRODUCTION

For new nonlinear devices it is important both to develop new materials as well a to investigate new methods to process and utilize more established materials.

It is of considerable practical interest to be able to frequency double GaAlAs lasers in order to realize small efficient light sources in the blue wavelength region. Waveguides in lithium niobate ($LiNbO_3$) is a promising approach. The main problem, however, is that the birefringence is not large enough to compensate for the wavelength dispersion in that case. Phase-matching can therefore not be obtained.

One way to circumvent this problem is to apply quasi-phase-matching by using a waveguide fabricated in such a way that the sign of the nonlinearity is periodically alternating along the guide. This technique is very versatile, in particular since any wavelength can be phase-matched by selecting an appropriate period for the modulation. For $LiNbO_3$, the nonlinear coefficient d_{33} can furhermore be used, which is 7 times larger than d_{31} which must be used when bire-fringency is utilized for phase-matching. However, a structure with alternating sign of the nonlinearity is very difficult to implement. Therefore quasi-phase-matching has very seldom been utlized in practice. Recently, however, fabricating techniques for such waveguides in $LiNbO_3$ have been reported by Webjörn, Laurell and Arvidsson (1989a, 1989b, 1989c) and by Lim, Fejer and Byer (1989a, 1989b). Reversal of the domain orientation in a surface layer on the positive c-face of c-cut $LiNbO_3$, that occur under certain heat treatments, has been utilized.

Quasi-phase-matched SHG in periodically poled polymeric waveguides has also very recently been reported by Khanarian et al (1989).

2. REVIEW OF EXPERIMENTAL RESULTS

Second harmonic generation of both green (Webjörn et al 1989a, Lim et al 1989a) and blue light (Webjörn et al 1989c, Lim et al 1989b) has been reported using periodically domain-in-verted waveguides in $LiNbO_3$. In the first experiments with a laser diode as pump 60 nW of blue light was generated from a fundamental power of 3 mW (Webjörn et al 1989c). The waveguides for QPM have been fabricated using different techniques. The ideal waveguide structure is illustrated in Fig 1. It has been found, however, that for the structures fabricated the boundary between regions of inverted and noninverted domain orientation tends to be triangularly shaped as illustrated in Fig 2 rather than rectangularly shaped. A promising fabrication technique (Webjörn et al 1989c) uses a periodic quartz pattern in combination with heat treatment up to 1100 °C to obtain a broad area with periodic domain reversal. Subsequently a channel waveguide is formed in this area using the proton-exchange process. The structure obtained is schematically illustrated in Fig 3. The angle γ has been estimated to be 30°. Since these waveguides are fabricated without titanium, the risk of photorefractive damage is minimized.

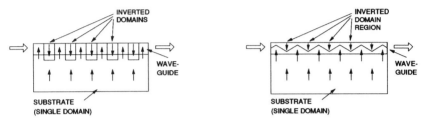

Fig 1. The ideal waveguide structure for quasi-phase matching in LiNbO₃. The waveguide passes regions with alternating ferroelectric domain orientation, corresponding to alternating sign of d₃₃, the largest nonlinear coefficient of LiNbO₃. By choosing the appropriate period an arbitrary wavelength can be frequency doubled.

Fig 2. Schematical illustration of actual waveguide structure used for quasi-phase matching (Webjörn et al, 1989c).

3. THEORY: INFLUENCE OF THE SHAPE OF THE DOMAIN BOUNDARY ON THE CONVERSION EFFICENCY

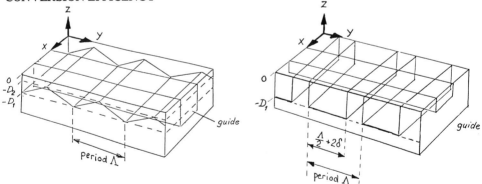

Fig 3. Waveguide geometry treated with triangularly shaped domain boundary.

Fig 4. Waveguide geometry treated,"laminar" case.

To calculate the conversion efficiency and investigate how it depends on the shape of the domain boundaries we used a theoretical model based on coupled mode theory and the effective index method. The geometries considered are illustrated in Figs 3 and 4. We assume both the fundamental and second harmonic (SH) light to be TM polarized, which is the relevant case for proton-exchanged waveguides in z-cut LiNbO₃. For low conversion efficiencies depletion of the pump at ω can be neglected and the slowly varying amplitude A of the mode generated at 2ω satisfies the equation:

$$\frac{\partial A(y)}{\partial y} = -i \cdot \frac{\omega}{2} \cdot \left(\frac{\mu_0}{\varepsilon_0}\right)^{\left(\frac{1}{2}\right)} \cdot N_\omega \cdot \int_{-\infty}^{\infty} \int_{-\infty}^{\infty} \frac{1}{n_{e2\omega}^2} \cdot h_{2\omega}^*(x,z) \cdot P_{NL}(x,y,z) \cdot e^{i\frac{2\pi}{\lambda}(N_{2\omega}-N_\omega)y} dx\,dz \qquad (1)$$

where N_ω and $N_{2\omega}$ are the effective indices for the pumping and generated second harmonic modes, respectively, $n_{e2\omega}$ is the extraordinary refractive index at 2ω, $h_{2\omega}$ is the transverse distribution of the SH mode, and λ is the second harmonic wavelength. P_{NL} is the complex amplidude of the nonlinear polarization induced by the pump, and described by:

$$P_{NL}(x,y,z) = A_0^2 \cdot d(y,z) \cdot e_\omega^2(x,z) = A_0^2 \cdot d(y,z) \cdot \left(\frac{N_\omega}{n_{e\omega}^2}\right)^2 \cdot \frac{\mu_0}{\varepsilon_0} \cdot h_\omega^2(x,z) \tag{2}$$

with e_ω and h_ω as z and x components of the transverse modal electric and magnetic fields at ω, respectively, A_0^2 the input pump power, $n_{e\omega}$ the extraordinary refractive index at ω. The nonlinear coefficient $d(y,z)$ varies spatially and is assumed to have reversed sign in the region(s) of inverted domain orientation. For the case of a triangular domain boundary we modelled the nonlinear coeficient with the following expression:

$$d(z,y) = \begin{cases} -d_{33} & -\cot\gamma \cdot (D_1+z) \leq y \pm m \cdot \Lambda \leq \cot\gamma \cdot (D_1+z) \\ +d_{33} & \text{otherwise} \end{cases} \tag{3}$$

where the angle γ and the depth D_1 of the domains are defined in Fig 3. For comparison we considered also the case of a laminar structure (rectangular domain boundaries) as shown in Fig 4.

Being a periodic function of y, $d(z,y)$ can also be expressed in terms of its spatial harmonics:

$$d(z,y) = d_{33} \cdot \sum_{n=-\infty}^{\infty} c_n(z) \cdot e^{-in\frac{2\pi}{\Lambda}y} \tag{4}$$

where for triangular domains we obtain:

$$c_n(z) = \frac{2}{\pi \cdot n} \sin\left\{n\frac{2\pi}{\Lambda}\cot\gamma \cdot (D_1+z)\right\} \qquad \text{for} \qquad -D_1 \leq z \leq -D_2 \tag{5a}$$

$$D_2 = \max\left[0,\left(D_1+\frac{T}{2}\cdot\tan\gamma\right)\right]4$$

and for rectangular domains:

$$c_n(z) = \frac{2}{\pi \cdot n} \sin\left\{n\frac{2\pi}{\Lambda}\left(\frac{\Lambda}{4}+\delta\right)\right\} \qquad \text{for} \qquad -D_1 \leq z \leq 0 \tag{5b}$$

The mismatch between the effective indices at the fundamental and the SH frequency occuring in the phase term of Eq (1) makes that term quickly oscillating. In order that the SH mode can build up constructively along the waveguide, the period Λ must be chosen so that one of the spatial harmonics of $d(z,y)$ compensates this phase mismatch. Comparing the exponential terms in Eqs (1) and (4) we find that this requirement is satisfied for:

$$\Lambda_n = \frac{n \cdot \lambda}{(N_{2\omega} - N_\omega)} \qquad ; \qquad n = 1,2,3... \tag{6}$$

where λ is the second harmonic wavelength. Neglecting the remaining asynchronous harmonics in Eq (4), substituting the resulting P_{NL} of Eq (2) into (1) and performing the integration with respect to y we obtain the expression for $A(L)$ and hence for the conversion efficiency η:

$$\eta = \frac{P_{2\omega}(L)}{P_\omega(0)} = d_{33}^2 \cdot L^2 \cdot P_\omega \cdot B^2 \cdot \left\{\int_{-\infty}^{\infty}\int_{-\infty}^{\infty} c_n(z) \cdot \frac{h_{2\omega}^* \cdot h_\omega^2}{n_{e2\omega}^2 \cdot n_{e\omega}^4} dx\,dz\right\}^2 \tag{7}$$

where $P_{2\omega}(L)$ is the power generated in the SH mode over the waveguide length L, and:

$$B = \frac{\omega}{2} \cdot \left(\frac{\mu_0}{\varepsilon_0}\right)^{\left(\frac{3}{2}\right)} \cdot N_\omega^3$$

The main difference compared to birefringence phase matching is the function $c_n(z)$, which is included in the overlap integral in Eq (7). For the case of a triangular boundary $c_n(z)$ is nonzero only within the layer definded by $-D_1 \leq z \leq -D_2$. The optimal case ("deep laminar structure") corresponds to rectangular boundaries (Fig 4) with $\delta=0$, D_1 much larger than the waveguide depth, and a period chosen so that phasematching is obtained for n=1. Eq (5c) then reduces to the constant factor $c_n(z) = 2/\pi$, which can be taken outside the integral sign. We thus have a reduction in the conversion efficiency of $(2/\pi)^2$, as reported earlier, eg by Somekh and Yariv (1972). For LiNbO$_3$, however, the influence of the larger nonlinear coefficent that can be used dominates and we have a net gain by $[(2/\pi)^2 \cdot (d_{33}/d_{31})^2] \approx 20$. To facilitate the fabrication it is possible to use a periodic structure with a period that is a multipel of the coherence length ($n=2,3,...$). For a deep laminar structures with $\delta = 0$ only odd values of n can be used. It should be noted however that for a triangular boundary and for rectangular boundaries with $\delta \neq 0$ also even values of n are allowed.

4. NUMERICAL RESULTS

Two examples with numerical results are given in Figs 5 and 6. The fundamental wavelength in the two cases is 1.064 μm and 0.83 μm, respectively. The calculations were carried out for graded index waveguides with refractive index increases that can be realized using annealed proton-exchanged waveguides. In both cases the waveguide was single mode for the fundamental wavelength. No systematic optimization of the waveguide parameters has been carried out. First order QPM is assumed (n=1) in both cases, and the angle γ is set to 30°. The period Λ is found to be 6.2 μm and 2.8 μm for the fundamental wavelength 1.06 μm and 0.83 μm, respectively. In Figs 5a and 6a the functions involved in the overlap integral are plotted. Figs 5b and 6b shows the conversion efficency as function of the depth D_1 for the triangular shaped boundary. A fundamental power of 100 mW and a waveguide length of 1 cm is assumed.

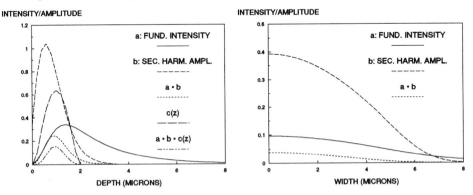

Figure 5a. Depth and width dependence for the functions included in the overlap integral of Eq (7). $c_n(z)$ is shown for the optimal position of the boundary in depth, i.e. for the optimal value of D_1. Fundamental wavelength 1.064 μm.

It is obvious from Figs 5b and 6b, that the conversion efficiency decreases considerably if the triangular domain boundary is not positioned at the optimal depth. This is probably the main reason why our first experimental conversion efficiencies have been lower than expected.

For the case of the fundamental wavelength 1.064 μm, the best conversion efficiency is ~10 times higher than for a corresponding bire-fringence phase-matched case (Fig 5b). The corresponding number for the wavelength 0.83 μm, is ~4 times higher (Fig 6b). The main reason for this smaller factor is the shorter period Λ in this case, which leads to a narrower range in depth over which the function $c_n(z)$ is nonzero (see Fig 6a). There is a tendency, however, that it is more easy to obtain a tight mode confinement for a shorter wavelength, which can partly compensate for this lower factor. Comparing the absolute values for the peak conversion efficiencies in Figs 5b and 6b, we find a slightly higher efficiency for the shorter wavelength in this example.

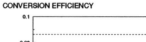
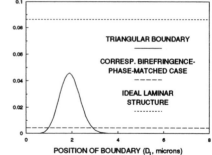

FUND. POWER: 0.1 WATT
LENGTH: 1 CM

Figure 5b. The conversion efficiency as function of the position of the boundary, and in comparison with an ideal laminar structure and a corresponding birefringence phase-matched case.

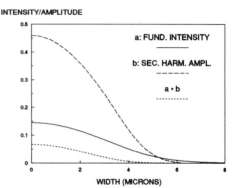

Figure 6a. As in Fig 5a but for a case with the fundamental wavelength 0.83 μm. Note the narrower range for which the function $c_n(z)$ is nonzero.

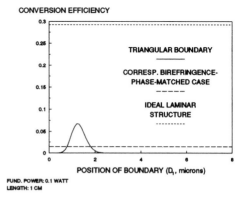

FUND. POWER: 0.1 WATT
LENGTH: 1 CM

Fig 6b. As in Fig 5b but for a case with the fundamental wavelength 0.83 μm.

5. CONCLUSIONS

It has recently been reported that periodically domain inverted waveguides have been fabricated in $LiNbO_3$ and used for quasi-phase-matched SHG. Blue light has been generated by frequency doubling a GaAlAs laser diode in such a waveguide. It has been found, however, that the boundary between inverted and noninverted regions tends to be triangular in shape rather than rectangular, which would be ideal. We have now derived an expression for the conversion efficiency in such waveguides, which can be used for optimizing the structure. It is concluded that in spite of the nonideal triangularly shaped domain boundary it should be possible to realize frequency doublers that are pumped with laser diodes and give output powers in the mW range.

6. REFERENCES

Khanarian G, Haas D, Keosian R, Karim D and Landi P   1989   "Phase Matched Second Harmonic Generation in a Polymeric Waveguide" *Conference on Lasers and Electro-optics (CLEO 89)*, Baltimore, USA, April 24-28, 1989, paper ThB1

Lim E J, Fejer M M and Byer R L   1989a   "Second-harmonic generation of green light in periodically poled planar lithium niobate waveguide" *Electron. Lett.* **25** pp 174-175

Lim E J, Fejer M M, Byer R L and Kozlovsky W J,   1989b   "Blue light generation by frequency doubling in periodically poled lithium niobate channel waveguide" *Electron. Lett.* **25** pp 731-732

Webjörn J, Laurell F and Arvidsson G   1989a   "Periodically domain-inverted lithium niobate channel waveguides for second harmonic generation" *Topical Meeting on Nonlinear Guided-Wave Phenomena: Physics and Applications*, OSA Technical Digest Series vol.2, pp 6-9

Webjörn J, Laurell F and Arvidsson G   1989b   "Fabrication of periodically domain-inverted channel waveguides in lithium niobate for second harmonic generation" accepted for publication in *J of Lightwave Technology* **7** (See also postdeadline paper PD3 OSA Topical Meeting on Integrated and Guided Wave Optics IGWO 89)

Webjörn J, Laurell F and Arvidsson G   1989c   "Laser diode light frequency doubled to blue in a lithium niobate channel waveguide" *Conference on Lasers and Electro-optics (CLEO 89)*, Baltimore, USA, April 24-28, 1989, Post-deadline paper PD10

Somekh S and Yariv A   1972   "Phase matching by periodic modulation of the nonlinear optical properties" *Opt Comm* **6** pp 301-303

Inst. Phys. Conf. Ser. No 103: Part 1
Paper presented at Int. Conf. Materials for Non-linear and Electro-optics, Cambridge, 1989

X-ray topographic study of alpha lithium iodate crystals under a static electric field

M. T. Sebastian*, H. Klapper* and S. Haussühl**

* Institut für Kristallographie, RWTH, D-5100 Aachen, Fed. Rep. Germany
** Institut für Kristallographie, Universität Köln, Fed. Rep. Germany

ABSTRACT: X-ray topographic studies are made on α-LiIO$_3$ crystals under a dc electric field along the polar hexagonal axis [001], which is the direction of high Li ion conductivity. The X-ray rocking curves recorded with insitu electric field show a strong enhancement of the diffracted intensity of the 0002 reflection whereas the hki0 reflections (i.e. reflecting planes parallel to the polar axis) do not show any intensity change. There are variations in the half-widths of the 0002 rocking curves for different applied voltages. The effect is present only when the electric field is applied along the polar axis. The 0002 X-ray topographs recorded with insitu electric field show extinction contrast consisting of striations parallel to the polar axis. This contrast and the enhancement of the 0002 intensity are attributed to lattice distortions by the space charge polarisation due to the movement of lithium ions under the influence of the electric field.

1. INTRODUCTION

Lithium iodate LiIO$_3$ exhibits a complicated polymorphism with at least four phases. Among them the α-phase belongs to the polar space group P6$_3$ and reveals high piezoelectric (Haussühl 1968; Warner et al 1970), non-linear optical (Nath, Haussühl 1969; Nash et al 1969), photoelastic (Warner et al 1970), pyroelectric (Bhalla 1984; Poprawski et al 1985) and electro-optic effects (Nash et al 1969). These interesting properties permit the production of various accousto-electronic and accousto-optical devices of high efficiency on substrates of these crystals (Avdienko et al 1977). The ionic conductivity is by a factor 10^3 higher along the polar axis [001] than normal to it, and it is also different for the [001] and [00$\bar{1}$] directions (Haussühl 1968; Holland-Moritz 1974; Remoissenet et al 1975). Several authors (Nagerl, Haussühl 1970; Holland-Moritz 1974; Remoissenet et al 1975; Lutze et al 1977) report that this conductivity is due to the motion of lithium ions through the structural [001] channels in the IO$_3$-framework (see fig.1). Practically no electrical conductivity occurs normal to [001]. Recent neutron and γ-ray diffraction studies (Qiang, Zhen 1988; Bouillot et al 1982; Li 1985; Hua-Chen 1981) of α-LiIO$_3$ crystals showed an enhancement of the diffracted intensity under an electric field. In the present paper we report the generation of a pronounced X-ray topographic extinction contrast by an electric field along the polar axis.

2. EXPERIMENTAL

The α–$LiIO_3$ crystals used in the present study were grown from aqueous solution by slow evaporation at $45°C$. Thin ($10\overline{1}0$) plates of about 0.3mm were cut with a string saw and polished on a wet cloth. A static electric dc field of 300 to 2000 V/cm was applied in different crystallographic directions using soft spongy graphite electrodes. X-ray topographs (Lang technique) and rocking curves (obtained by stepwise rotation of the crystal on the Lang camera) were recorded using AgKα–radiation (linear absorption coefficient $\mu_o = 6.0$ mm^{-1}).

3. RESULTS AND DISCUSSION

X-ray rocking curves recorded from all crystals showed a strong enhancement of the diffracted intensity of the 0002 reflection under an electric field along the polar axis. This is in agreement with earlier reports (Bouillot et al 1982; Quiang, Zhen 1988). Fig.2 shows a typical rocking curve. The intensity increases instantaneously on applying the field and reaches a saturation after a few minutes. After removal of the field the intensity decreases slowly within about one hour and comes to a level which is higher than the original intensity without field. The peak intensity depends strongly on the field strength (fig. 2). No increase, however, occurs at fields beyond 1500V/cm.
There is a slight shift in the peak maxima towards the Kα_2 position which may be due to the merging of the Kα_1 and Kα_2 lines at high voltages. This results from the increase of the rocking curve width with increasing field, i.e. the electric field increases the width of reflection range and thus reduces the primary extinction.

All 0002 X-ray topographs taken with insitu electric field along the polar axis [001] show a strong kinematical (extinction) contrast consisting of striations parallel [001] (see figs. 3 and 4 for two different crystals). The striations disappear on removal of the electric field as evidenced by figs. 3f and 4e which were recorded immediately after 3e and 4d respectively. The striations are present only when the electric field is parallel to the polar axis. An electric field of 650 V/cm normal to [001] did not produce any contrast (see fig. 3c). The contrast is present only if reflections with l ≠ 0 are used for imaging, it is invisible for reflections hki0. This proves that the atomic displacements in the (electrically induced) strain field are esentially parallel to [001].

The application of a high electric field for a long time leads to cracking of the crystals along the ($10\overline{1}2$) faces (labels C in figs. 3,4). Along the cracks a white powder develops which could be Li_2CO_3 (Holland-Moritz, 1974). It is found that during long-time solid state electrolysis lithium iodate decomposes at both electrodes. This can also be recognized in figs. 3,4. A detailed study of the $LiIO_3$ decomposition at both electrodes during electrolysis under various conditions is presented by Holland-Moritz (1974). In the presence of hydrogen, conduction by protons occurs.

The X-ray topographic striation contrast is, after a certain relaxation period, reversibly connected with the electric field. Thus it cannot be attributed to dislocations. From observation of various specimens it is found that the striations are mostly non-uniform in distribution and 'intensity'. A significant and systematic dependence on the direction of the field is not observed. Frequently the striation start or end at growth

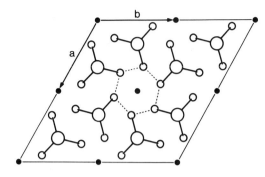

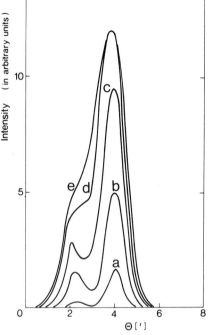

Fig. 1. Structure of α-LiIO$_3$ projected along the polar axis. Large and small open circles: Iodine and oxygen. Small full circles: Lithium. Dotted line: Channel in the IO$_3$-framework. a = 5.48 Å.

Fig. 2. X-ray 0002 rocking curves of α-LiIO$_3$, taken with an insitu dc electric field along the polar axis. (a) Short-circuited. (b) 350 V/cm, (c) 700 V/cm, (d) 1500 V/cm, (e) 2100 V/cm.

bands or growth sector boundaries (e.g. label X in fig. 3a,d).

The increase and decay of the diffracted 0002 intensity cannot be due to a mere piezoelectric (electrostrictive) interaction. It obviously arises from the movement of the lithium ions along the [001] channels of the IO$_3$-framework. Although details are not known it is assumed that the movement of ions along [001] (note that no effect is observed for a field normal to [001]) leads to an inhomogeneous distribution of positive charges within the (negative) IO$_3$-framework distorting the crystal lattice. It is noteworthy that the displacements generating the lattice distortions are directed parallel to the polar axis.

The start or stop of the striation contrast at growth bands or growth sector boundaries indicate that the ionic conductivity is blocked at such growth defects, thus leading to higher local charge accumulations and stronger X-ray topographic contrast than elsewhere.

Optical contrast lines corresponding to the above X-ray topographic striations have been observed by Li (1985) using a polarizing microscope. Li found bright thin columns sprouting from the cathode and extending along [001] towards the anode. They are assumed to result from the accumulation of interstitial lithium ions in bundles of conducting channels. Zhang et al (1981) and Li (1985) report that the potential difference across the crystal is steep on the anode as well as at the cathode side and rather flat in the middle. This indicates an accumulation of positive charges (Li$^+$

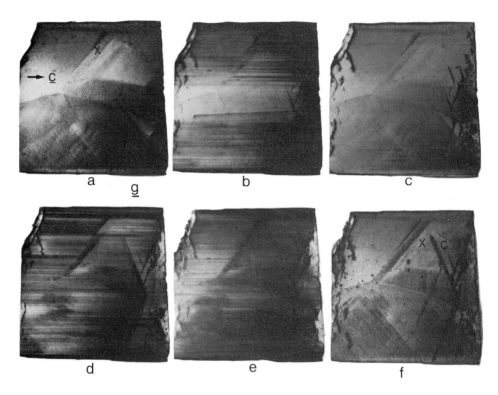

Fig. 3. 0002 X-ray transmission topographs of a (10$\bar{1}$0) plate (about 15 x 14 mm). (a) Short–circuited. (b) –300 V/cm along [001]. (c) +650 V/cm normal to [001]. (d) +650 V/cm along [001]. (e) –1300 V/cm along [001]. (f) Again short–circuited.

ions) at the cathode side of the crystal and of negative charges (Li$^+$ vacancies) at the anode side. Consequently the electric field must be much lower in the middle of the probe than in its both ends close to the electrodes. No significant evidence of this field distribution appears on the X-ray topographs, though in some cases a higher diffracted intensity is observed at the anode side of the probe (fig. 3b).

An increase of the neutron 0002 reflection intensity is also induced by an electric ac field along the polar axis (Hua–Chen 1981). The magnitude of the enhancement varies inversely with the ac frequency and disappears at about 1 kc/sec.

In conclusion it is noted, that an enhancement of diffracted intensity and the appearance of X–ray topographic striation contrast by the application of an electric field also has been observed in quartz (conduction of impurity ions, Calamiotou et al 1987; Sebastian et al 1988) and in lithium hydrazinium sulphate (proton conductor, Sebastian and Klapper 1989). In quartz the effect is due to the separation of the charge–compensating alkali ions under the action of the electric field. The topographic contrast disappears on X–ray irradiation of the crystal during which Al hole centres are formed, neutralizing the negative space charges generated by electrolysis.

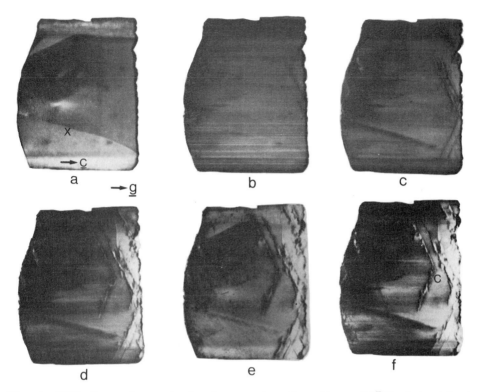

Fig. 4. 0002 X-ray transmission topographs of another (10$\bar{1}$0) plate (about 8 x 9 mm) with dc field along [001]. (a) Short–circuited. (b) +500 V/cm. (c) +1000 V/cm. (d) +1500 V/cm. (e) Short–circuited. (f) –1500 V/cm.

ACKNOWLEDGEMENTS

One of the authors (MTS) is grateful to the Alexander von Humboldt Foundation for the award of a fellowship. We wish to thank Mr. R. A. Becker for making a high voltage X-ray topographic chamber and to Mrs. Delmhorst for reproducing the photographs.

REFERENCES

Avdienko K I, Bogdanov S V, Kidyarov V I, Semyonov and Sheloput D V (1977) Izv. Akad. Nauk. Ser. Fiz. **4** 700

Bhalla (1984) J. Appl. Phys. **55** 1229

Bouillot J, Baruchel J, Remoissenet M, Joffrin J and Lajzerowicz J (1982) J. Physique **43** 1259

Calamiotou M, Psicharis V, Filippakis S E and Anastassakis E(1987) J. Phys. **C 20** 5641

Haussühl S (1968) Phys. Stat. Sol. **29** K159

Holland-Moritz I (1974) Thesis, Universität Köln

Hua-Chen Hu (1981) J. Phys. Lett. **42** L189

Li Yinyuan (1985) Advances in the Science of China. Physics Vol. **1**. Science Press Beijing.

Lutze R, Gieseka W and Schroter W (1977) Sol. St. Commun. **23** 215

Nash F, Bergman J G, Boyd G D and Turner E H (1969) J. Appl. Phys. **40** 13

Nath G and Haussühl S (1969) Appl. Phys. Letters **14** 154

Nagerl H and Haussühl S (1970) Phys. Stat. Sol. **a3** K203

Qiang Li and Zhen Yang (1988) Z. Kristallogr. **183** 265

Remoissenet M, Garandel J and Arend H (1975) Mater. Res. Bull. **10** 181

Sebastian M T and Klapper H (1989) 12th Europ. Crystallogr. Meeting Moscow

Sebastian M T, Zarka A and Cappelle B (1988) J. Appl. Cryst. **21** 326

Warner A W, Pinnow D A, Bergman J G and Grane G R (1970) J. Accoust. Soc. Amer. **47** 791

Zhang et al (1981) Chinese Physics (AIP) **1** 923

From atomic positions to nonlinear optical properties in materials of the KTiOPO$_4$ (KTP) family of second-harmonic generators

P.A.Thomas
Clarendon Laboratory, Parks Road, Oxford OX1 3PU, U.K.

ABSTRACT: Calculations of the refractive indices of KTiOPO$_4$ using point-dipole theory are presented and interpreted within the existing model for the structural origin of its high non-linear optical response. It is shown that interactions in the Ti-O system are of principal importance to the calculation of the linear susceptibilities and especially that the oxygen atoms involved in the anomalously short bonds to the must be given heightened polarizability contributions for the correct modelling of both magnitude and sign of the optical anisotropy. Bond polarizability theory is used to discuss the nonlinear optical response of KTiOPO$_4$ and its Rb and Sn analogues.

1. INTRODUCTION

Potassium titanyl phosphate, KTiOPO$_4$ (KTP), orthorhombic (space group Pna2$_1$, C$_{2v}^9$) has been of interest for some years because of its strong nonlinear optical properties (Zumsteg, Bierlein and Gier 1976). KTP is a particularly efficient doubler of the Nd:YAG fundamental frequency (1.06μ) into the green part of the optical spectrum and is an advantageous material because of its high laser-damage threshold, large temperature band-width for phase-matching and its mechanical robustness in addition to its high basic nonlinearity. The principal structural origin of the large optical nonlinearity in KTP has long been considered to be the anomalously short Ti-O bonds which give rise to highly distorted TiO$_6$ octahedra (Tordjman, Masse and Guitel 1974). KTP is also capable of several compositional modifications e.g. K may be substituted by Rb, Tl (Masse and Grenier 1971) or Na (Bamberger, Begun and Cavin 1988) and P by As (Brahimi and Durand 1986). Therefore, the KTP family provides an ideal system for the systematic study of the relationship between crystal structure and optical properties.

2. DISCUSSION

The modelling of the linear optical susceptibilities of KTP is important because the interactions which dominate the linear optical properties are usually those which dominate the nonlinear properties also. This relationship is expressed through the Miller (1964) rule which may be written in the form appropriate to second-harmonic generation as

$$\Delta_{ijk} = d_{ijk} / (\chi_i (2\omega)\chi_j (\omega)\chi_k (\omega)) \qquad \text{--------(1)}$$

where d_{ijk} is a component of the second-order susceptibility tensor, $\chi_i (2\omega)$ is the linear susceptibility at the doubled frequency, and $\chi_j (\omega)$ and $\chi_k (\omega)$ are the appropriate linear susceptibilities at the fundamental frequency. Miller found that the values taken by the Δ's of materials showing widely differing d_{ijk}'s were of the same order

The relative constancy of the Miller Δ ratio thus expresses the important relationship which exists between the linear and nonlinear susceptibilities.

Polarizability theory is well-known as a method for the calculation of optical properties of crystals. In recent years, the point-dipole model in which each atom of the crystal structure is represented by a point-dipole oscillator having a specified isotropic polarizability, has been used successfully for the calculation of optical activity and refractive indices in a number of inorganic compounds (Devarajan and Glazer 1986, Thomas 1988). It is usually found that in addition to obtaining calculated values of the optical parameters in agreement with experiment, valuable insight into the origin of optical effects is gained. Point-dipole theory differs from bond-polarizability theory because the optical transitions are considered implicitly to be *intra-ionic* i.e. the response of each individual atom/dipole to the propagating field is calculated and then all responses are summed to give the total macroscopic optical polarization. In bond polarizability theory, the response of the bonding electrons to the propagating field is considered and therefore the interactions are implicitly *inter-ionic*. This approach has a clear affinity to the band-theory picture of solids and leads to the appearance of optical anisotropy (birefringence) in a natural way as pointed out by Weber (1988). However, there is insufficient experimental data currently available to allow bond polarizability calculations of the refractive indices of complex crystals such as KTP.

The experimental refractive index data at 6328 $\overset{\circ}{A}$ are taken from Fan, Huang, Hu, Eckardt, Byer and Feigelson (1987). They are:-

$$n_x = 1.7626 \quad n_y = 1.7719 \quad n_z = 1.8655$$
$$\Delta n_{xy} = 0.0093 \quad \Delta n_{yz} = 0.0936 \quad \Delta n_{xz} = 0.1029$$

Initially, we must decide on input polarizability volumes for the atoms of the structure. The values we assign reflect the importance of each atom to the optical properties and, because the calculated influence of neighbouring atoms is used in making the polarizabilities anisotropic, inter-atomic interactions are reflected also. In KTP, we have 4 types of atoms (K, Ti, P and O) and 16 independent atomic sites. We could, if we so chose, decide to give each of the 16 independent atoms different polarizability volumes thus distinguishing between O(1) and O(4), K(1) and K(2) etc. However, we have only three experimental parameters (n_x, n_y and n_z) against which to measure the fit and therefore choices must be made about which atomic polarizabilties should be varied. In terms of polarizability theory alone we expect the contributions to decrease in the order α_O, α_K, α_{Ti}, α_P. As a reasonable approximation, we set α_P to a negligibly small value which is not varied. Therefore, we are concentrating on the TiO_6 octahedra and the KO_8 and KO_9 cages. The effect of the contact between a K atom and an O atom is much smaller than that between a Ti atom and an O atom because the K-O bond lengths are much longer (average K-O contact length = 2.889 $\overset{\circ}{A}$ average Ti-O contact length = 1.969 $\overset{\circ}{A}$)

The major trends found from the calculations are :-
(i) The optical anisotropy cannot be modelled correctly using a single value of α_O - the birefringences Δn_{yz} and Δn_{xz} are far too low and the **sign** of the birefringence Δn_{xy} in the plane perpendicular to the polar axis is reversed. The polarizability ellipsoids of the O atoms are not very anisotropic whereas high oxygen anisotropy is expected from a highly birefringent material such as KTP.

(ii) The results of the calculations immediately improve when the two O atoms, denoted by OT1 and OT2, involved in the *anomalously short* bonds to Ti atoms are treated differently and are given **higher** polarizabilities. The value and sign of Δn_{xy} is found to be critically dependent on the ratio α_{OT} : α_O (Figure 1). The polarizability ellipsoids for the OT atoms become markedly more anisotropic with particular extension in the polar direction (parallel **c**) as shown by the contrast between Figures 2a and 2b . The refractive index parallel to **C** tends to be too high because the extension of the OT ellipsoids is too extreme.

Figure 1 (right) shows the calculated birefringence Δn_{xy} plotted as a function of the ratio of α_{OT} and α_O polarizability volumes. Note that the calculated Δn_{xy} undergoes a change of sign over the range of values for the ratio and that it passes through a uniaxial position at X. The inset shows the variation within the ratio range 1.54-1.59 as we approach the experimental value of Δ (0.0093) at Y.

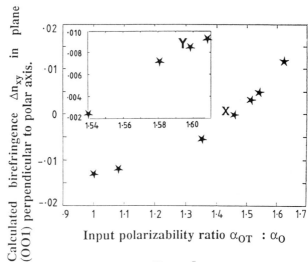

Figure 1

Figure 2a (left) shows a view of the polarizability ellipsoids for the Ti atoms (x5) and the OT atoms. The figure corresponds to a single input oxygen polarizability (1.85 Å³). The birefringences n_{xz} and n_{yz} calculated are much too low (0.0242 and 0.0372 respectively) and Δn_{xy} has the wrong sign.

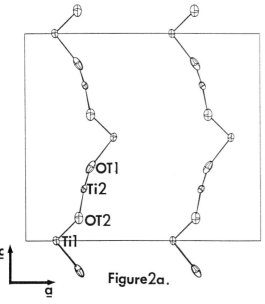

OT1
Ti2
OT2
Ti1

Figure 2a.

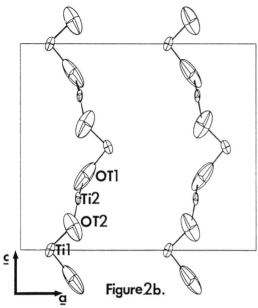

Figure 2b (left) shows the same view of the structure as figure 2a and is drawn to the same scale. OT and O now take polarizabilities corresponding to point **Y** on figure 1. The birefringences Δn_{xz} and Δn_{yz} are much closer to the experimental values (in fact, at 0.1159 and 0.1069 respectively, they are slightly high) and Δn_{xy} has the correct sign and magnitude.

Figure 2b.

(iii) Inclusion of the K atoms with a reasonable polarizability volume acts to bring down n_z while affecting Δn_{xy} very little. Therefore, the results can be "fine-tuned" using the K atom contribution. The K atoms are also bonded to the OT atoms (Figure(3)) and therefore exert a second-order influence on the polarizability ellipsoids. Note that the major components of the K atoms' polarizability ellipsoids are directed perpendicular to the Ti1-OT2-Ti2-OT1... chains in the structure and so they will tend to fatten the OT ellipsoids perpendicular to the polar axis and to bring the anisotropies in the (010) and (100) planes down. This effect on the OT ellipsoids can be seen in Figure 3 in comparison with Figure 2b.

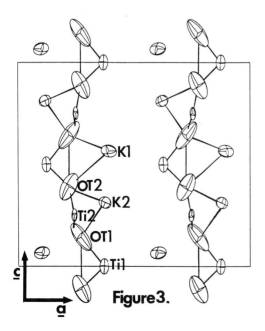

Figure 3 (left) shows a view along [010] of the polarizability ellipsoids for the Ti, OT and K atoms in the KTP structure. The sizes of the K and Ti ellipsoids have been scaled up by a factor of 5. The input polarizability volumes were

$$\alpha_K = 0.51 \text{ Å}^3$$

$$\alpha_{Ti} = 0.1822 \text{ Å}^3$$

$\alpha_O = 1.6347 \text{ Å}^3$ and $\alpha_{OT} = 2.8254$ Å^3. The calculated refractive index values were $n_x = 1.7540$ $n_y = 1.7645$ and $n_z = 1.8638$. The achievement of birefringences in the (010) and (100) planes close to experimental values without loss of anisotropy in the (001) plane is brought about through the tuning of the α_K value.

Figure 3.

3. NONLINEAR OPTICAL PROPERTIES

In the preceding discussion it has been shown that the optical anisotropy of KTP is dominated by Ti-OT-O interactions and influenced to a lesser extent by K-O interactions. The use of bond polarizability theory in calculating nonlinear optical (NLO) effects again requires that we assess the dominant interactions in the system this time by assigning polarizability contributions to the bonds in the crystal structure according to certain criteria (Levine 1973 and others). The contributions of the bonds are then summed over the unit cell taking the structural arrangement into account through geometrical factors, to give the total contribution to a single d_{ijk} component. Hansen, Protas and Marnier (1988) used the principles of the theory semi-empirically to extract bond polarizability for three types of bond in the KTP structure i.e., the two anomalously short bonds (Ti1-OT2 and Ti2-OT1), the medium Ti-O bonds (four per TiO_6 octahedron) and the P-O bonds, from the measured values of the d_{ijk} coefficients. Their essential assumption was that the short Ti-OT bonds had to be treated distinctly and that, in comparison with these, the contribution of the long Ti-O bonds (Ti1-O1 and Ti2-OT1) could be neglected and the remaining 8 Ti-O bonds treated as the same bond species. The treatment resulted in a derived bond-polarizability value for the short Ti-O bonds of approximately 14 x the medium bonds' value.

We have grown crystals of KTP, $RbTiOPO_4$ (RTP) and $KSnOPO_4$ (KSP) and used single-crystal X-ray diffraction to find the atomic positions in the crystal structures. We have then measured the second-harmonic generation (SHG) of comparable powder samples of all three using a Nd:YAG laser as the fundamental. Given KTP as a standard, the results are as follows:-

RTP output intensity = 85 % KTP ouput intensity
KSP output intensity = 2% KTP output intensity.

A brief explanation of these observations is offered here with reference to the bond-lengths in the MO_6 (M=Ti/Sn) octahedra (Table (I)).

Table (1)

Bond lengths in Ti and Sn oxygen octahedra in KTP, RTP and KSP.				
Bond	KTP	RTP	Bond	KSP
Ti1-O1	2.150	2.148	Sn1-O1	2.091
Ti1-O2	1.958	1.949	Sn2-O2	2.093
Ti1-OT1	1.981	1.952	Sn1-OS1	1.978
Ti1-OT2	1.716	1.717	Sn1-OS2	1.975
Ti1-O5	2.043	2.072	Sn1-O5	2.112
Ti1-O6	1.987	2.016	Sn1-O6	2.063
Ti2-O3	2.044	2.061	Sn2-O3	2.134
Ti2-O4	1.981	2.005	Sn2-O4	2.102
Ti2-OT1	1.733	1.746	Sn2-OS1	1.957
Ti2-OT2	2.092	2.089	Sn2-OS2	1.961
Ti2-O7	1.964	1.971	Sn2-O7	2.052
Ti2-O8	1.991	1.995	Sn2-O8	2.076

Mean Ti-O bond lengths in KTP:- 1.972 Å with s.d. 0.143 Å and 1.967 Å with s.d. 0.122Å.
Mean Ti-O bond lengths in RTP:- 1.976 Å with s.d. 0.134 Å and 1.978 Å with s.d. 0.122Å.
Mean Sn-O bond lengths in KSP:- 2.052 Å with s.d. 0.06Å and 2.047 Å with s.d. 0.07Å.

The Ti-O bond-lengths in KTP and RTP are very similar and the level of the distortions of the octahedra in both compounds which is reflected in the standard deviations on the mean Ti-O bond-lengths, is about the same. Therefore, if the Ti-O interactions dominate the NLO response, we expect KTP and RTP to perform at the same level. The substitution of K by Rb increases both the linear refractive indices and the unit cell volume and therefore acts to lower the output SHG power in the powder test. (Note that K-O and Rb-O interactions are not otherwise brought into the picture of the NLO response - their most important influence is through their effect on the linear susceptibilities.) In KSP however, the Sn-O bond lengths are equalized and the octahedra are very much less distorted. In fact, KSP is very close to a centrosymmetric structure which we have identified as the likely high-temperature prototype structure for materials of the KTP family. In the SnO_6 octahedra, severely shortened Sn-O bonds equivalent to the short Ti-O bonds in KTP and RTP do not appear. Thus, although KSP is in all other ways a complete structural analogue of KTP, the absence of the short bonds and high MO_6 distortions is sufficient to depress its SHG output by more than an order of magnitude. These observations, which are the subject of a fuller paper currently in preparation for submission to J.Phys.C, prove that the short Ti-O bonds are the features essential to the strong NLO properties of KTP.

4. REFERENCES

Bamberger C E, Begun G M and Cavin O B 1988 J.Solid State Chem. **73** 317

Devarajan V and Glazer A M 1986 Acta Crystallogr. A42 560

El Brahimi M and Durand J 1986 Rev. de Chimie Minerale **23** 146

Fan Tso Yee, Huang C E, Hu B Q, Eckardt R C, Fan Y X, Byer R L and Feigelson
    R S 1987 Appl. Optics **26**(12) 2390

Hansen N K, Protas J and Marnier G 1988 C.R.Acad. Sci. Paris **307**(II) 475

Levine B F 1973 Phys. Rev. B **7**(6) 2600

Masse R and Grenier J 1971 Bull. Soc. fr. Mineral. Crystallogr. **94** 437

Miller Robert C 1964 Appl. Phys. Lett. **5**(1) 17

Thomas P A 1988 J. Phys. C. **21** 4611

Tordjman I, Masse R and Guitel J C 1974 Z.Krist. **139** 103

Weber Hans-J 1988 Acta Cryst. A**44** 320

Inst. Phys. Conf. Ser. No 103: Part 1
Paper presented at Int. Conf. Materials for Non-linear and Electro-optics, Cambridge, 1989   65

New non-linear optical materials of the KTiOPO$_4$ family

J. Mangin, G. Marnier, B. Boulanger and B. Menaert

Laboratoire de Minéralogie, Cristallographie et Physique Infrarouge
URA - CNRS n° 809 - Université de Nancy I
B.P. 239 - 54506 Vandoeuvre les Nancy Cedex - France

ABSTRACT : Millimetric crystals of new compounds of the KTP type have been obtained by the flux growth method. The general formula is $Cs_xM_{1-x}TiOAsO_4$ with M = K or Rb. A powder technique for second harmonic generation of the Nd : YAG laser radiation at 1.064 µm is used to assess the non-linear behaviour and to determine the ferroelectric transition temperatures. The range of transparency is examined on four compounds; as in KTiOPO$_4$, the extinction reaches 100 % at 0.35 µm for 1 mm thick samples. The clear transmission range extends from 0.35 µm to about 5 µm, where the beginning of infrared absorption is probably due to molecular AsO$_4$ absorption bands.

1. INTRODUCTION

Since early works of Bierlein (1976) and Zumsteg et al.(1976) on hydro-thermally grown crystals, KTiOPO$_4$ (KTP) has been proved to be a very at-tractive material for non-linear optical and electro-optical applications. An extensive litterature survey and update of its properties and applica-tions may be found in a recent paper of Bierlein and Vanherzeele (1989). Among several isotypes of KTP already elaborated, a promising alternative has been also proposed by Bierlein et al. (1989) with KTiOAsO$_4$ (KTA). In our laboratory the flux growth method developed by Marnier (1986), along with following improvements (Marnier et al. 1987), has allowed to synthe-tize new non-linear materials of the KTP family with the general formula $Cs_xM_{1-x}TiOAsO_4$, M representing K or Rb. Optically clear crystals were ob-tained with size varying from hundred microns up to a few millimeters. The chemical composition of these solid-solutions was established by electro-nic microprobe analysis.

As a preliminary investigation on the optical properties, this work was carried out to determine the effective non-linear optical behaviour and the wavelength range of transparency of these compounds. This study en-able also to determine the melting and ferroelectric transition tempera-tures.

2. OPTICAL TRANSMISSION

In the following the abbreviation TA will represent the titanyl arsenait. TiOAsO$_4$ group. Four single crystals were selected for meeting requirements of size, optical quality and chemical compositions : $Rb_{0.91}Cs_{0.09}TA$, $Cs_{0.84}K_{0.16}TA$, $Cs_{0.31}K_{0.69}TA$ and $Cs_{0.09}K_{0.91}TA$. The first one is close to

the RbTA pole and the last one, close to the KTA pole, was slightly yellow. The second solid solution was the nearest to the CTA pole. The third
one is an intermediate case.

The transmission of plane parallel plates of unoriented samples (thickness
1 mm) was obtained from 0.2 μm to 40 μm on Perkin-Elmer 340 and 457 spec-
trophotometers. The results are drawn on figure 1.

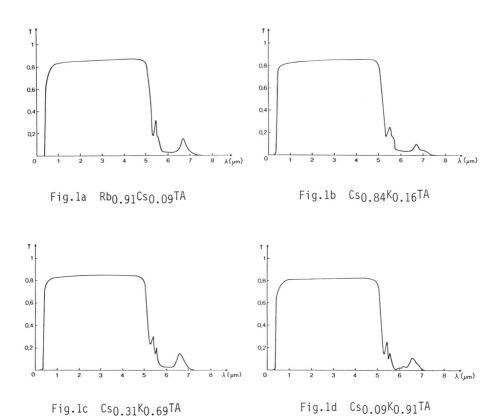

Fig.1a $Rb_{0.91}Cs_{0.09}TA$ Fig.1b $Cs_{0.84}K_{0.16}TA$

Fig.1c $Cs_{0.31}K_{0.69}TA$ Fig.1d $Cs_{0.09}K_{0.91}TA$

Fig.1   Transmission spectra of $Cs_xK_{1-x}TiOAsO_4$ solid solutions.

All the samples exhibit the same range of transparency : the extinction
reaches 100 % at 0.35 μm and the infrared absorption begins at 5 μm which
is 10 % higher than the value reported by Bierlein et al. (1989) for KTA.
Similar spectra performed on KTP showed the occurence of a first strong
absorption peak at λ = 4.75 μm. Taking only into account the mass difference between the (AsO_4) and (PO_4) tetrahedra, this would lead to a first
absorption peak at 5.7 μm for KTA, close to the observed value of 5.32 μm
on fig.1. This show that the near infrared absorption is probably due to
molecular $(PO_4)/(AsO_4)$ absorption bands.

3.   FERROELECTIRC TRANSITION TEMPERATURE

X-Ray diffraction data of Protas et al. (1989) showed that CTA belongs to
the acentric point group mm2, the space group being $Pna2_1$ at room tempe-

rature. Likewise KTP (Tordjman et al. 1974), its isotypes KTA, RbTA and TlTA were shown to belong to the same point group by El Brahimi and Durand (1986).

Yanovskii and Voronkova (1986) studied dielectric permeability and second harmonic generation (SHG) intensity variations with temperature on powdered samples of KTP, RbTP and TlTP. The two methods gave similar determinations of the ferroelectric transition temperatures : the SHG vanishes with the change of symmetry $Pna2_1$ (acentric) to Pnam (centric), while a sharp anomaly of the dielectric permeability ε_{33} is observed.

Figure 2 gives the SHG vanishing temperatures of five examines solid solutions; test measurements performed on KTP and RbTP powders gave the same results as those of Yanovskii and Voronkova (1986). We observed a continuous decrease of the number of grains in phase-matching conditions with the increase of temperature. Variations with the temperature of the refractive indices of the fundamental and harmonic frequencies may be different and especially near the ferroelectric phase transition because of the sharp anomaly of the dielectric permeability ε_{33}. Figure 2 exhibits a non linear decrease of the transition temperature with the rise of alkaline cesium concentration.

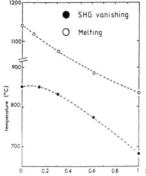

Fig.2 Melting and ferroelectric transition temperatures of $Cs_xM_{1-x}TiOAsO_4$ solid solutions.

Also plotted on figure 2 are the melting points of the compounds; the measurements were performed on millimetric crystals, watched through a binocular lens during heating. All the samples become opaque at about 150° C before their melting point, due to chemical decomposition, leading to an estimated error of ± 10° C on the given values.

4. CONCLUSION

Optical spectra of new-non linear solid solutions of general formula $Cs_xM_{1-x}TA$ (M = K or Rb) showed a clear transmission range extending from 0.35 µm to 5 µm; beyond 5 µm, the infrared absorption is due to molecular (AsO_4) absorption bands. SHG vanishing measurements as a function of temperature exhibit a non linear decrease of the ferroelectric transition temperature with the rise of cesium concentration. The melting temperature of KTA and CTA poles are respectively 1140° C and 970° C, ± 10° C. Also analyzed, the RbTA pole gave a SHG vanishing at 794° C and a melting temperature of 1125° C.

5. REFERENCES

Bierlein J D and Gier T A 1976 US patent n°3, 949, 323

Bierlein J D and Vanherzeele H 1989 J. Opt. Soc. Am. B 6, 622

Bierlein J D, Vanherzeele H and Ballman A A 1989 Appl. Phys. Lett. 54, 783

El Brahimi M and Durand J 1986 Rev de Chimie Minérale 23, 147

Marnier G 1986 CNRS French Patent n° 2, 609, 976

Marnier G, Boulanger B, Menaert B and Metzger M 1987 CNRS French Patent
   n° 8, 700, 811

Protas J, Marnier G, Boulanger B and Meneart B 1989 Acta Cryst. (to be
   published)

Tordjman I, Masse R and Guitel J C 1974 Z. Krist. 139, 103

Yanovskii V K and Voronkova V I 1986 Phys. Status Solidi A 93, 665

Zumsteg F C, Bierlein J.D. and Gier T A 1976 J. Appl. Phys. 47, 4980

Inst. Phys. Conf. Ser. No 103: Part 1
Paper presented at Int. Conf. Materials for Non-linear and Electro-optics, Cambridge, 1989

69

Room temperature electronic optical nonlinearities in polycrystalline ZnSe

W Ji, J R Milward, A K Kar, B S Wherrett and C R Pidgeon

Department of Physics, Heriot-Watt University, Edinburgh EH14 4AS

ABSTRACT: We have observed self-defocussing of nanosecond laser pulses in ZnSe near the band gap. The refractive cross-sections have been calculated from this in the wavelength range 470-476 nm. We have also carried out an experiment on degenerate four-wave mixing to calculate the effective nonlinear cross-sections in the same spectral range. The refractive cross-section has been measured to be as large as $\sim 5 \times 10^{-20}$ cm^3. Comparison with the three level band-filling theory indicates that the contribution from the dynamic Burstein-Moss shift dominates the refractive cross-section.

1. INTRODUCTION

Over the last few years there has been increasing interest in the optical nonlinearities of ZnSe because of its potential in the application of nonlinear optical devices for optical signal processing and optical computing. Thermal nonlinearities in ZnSe have been investigated extensively. Optical bistability by increasing absorption has been observed in waveguides by Kim et al. (1987) and in bulk structure by Kar & Wherrett (1986) and Taghizadeh et al. (1985). Thermally refractive nonlinearities have been employed as the mechanism for optical bistability in interference filters by Smith et al. (1984), Olbright et al. (1984) and Wherrett and Smith (1986). Two-photon absorption and the subsequent photogenerated free-carrier defocussing in ZnSe have been studied using picosecond laser pulses by Van Stryland et al. (1985). Passive optical limiters based on such nonlinearities have been proposed and demonstrated by Van Stryland et al. (1988). All of these studies have been conducted at wavelengths well below the band edge of crystalline ZnSe. Recently exciton saturation in molecular beam epitaxy (MBE) grown ZnSe thin films has been reported and the corresponding refractive nonlinearity has been calculated by Peyghambarian et al. (1988).

In this paper we report the observation of two nonlinear effects of electronic origin in ZnSe: self-defocussing and degenerate four-wave mixing. These effects result from the photoexcitation of free carriers by nanosecond laser pulses at wavelengths just below the band edge. The transmitted spatial profiles and phase conjugate signals have been measured in the self-defocussing and DFWM experiments respectively. The refractive cross-section, σ_n, which is defined as the refractive index change per electron-hole pair in unit volume, is measured to be as large as -5×10^{-20} cm^3. This value is comparable to that inferred for MBE grown ZnSe thin films by Peyghambarian et al. (1988) and observed for the narrow gap material GaAs by Lee et al. (1988). To our knowledge, this is the first report of such measurements in room-temperature ZnSe near the band edge.

In section 2 we report the observation of self-defocussing in ZnSe. We present a theoretical model to describe this phenomenon, and fits to the measured spatial profiles yield the spectral response of the nonlinear refractive cross-section. In section 3 we describe the degenerate four wave mixing experiment. With modification, we apply the theory of Wherrett et al. (1983) to obtain the effective nonlinear cross-section, $\sigma_{\it eff}$. In section 4 we compare the measured cross-sections with the bandfilling theory.

2.   SELF-DEFOCUSSING

Zinc selenide is a direct gap semiconductor with a relatively large band-gap energy of 2.67 eV at 300 K - Landolt-Bornstein (1982). As we tuned the wavelength of the laser pulses towards the vicinity of the band edge, a number of nonlinear phenomena were observed. One of these effects was self-defocussing. In our experiments, the tunable laser radiation was provided by an excimer pumped dye laser system which produces 5 ns (FWHM) pulses with a maximum pulse energy of 0.5 mJ. The experimental set-up is shown in Fig. 1. The experiments were conducted at a pulse repetition rate of 2 Hz to minimise thermal effects. The pulses were spatially filtered and focussed down onto the 212 μm thick, CVD grown, polycrystalline ZnSe sample. The sample was placed ~ 6 cm (z_o) behind the beam waist (90 μm HW e^{-2} M) to give a spot size on the sample of 170 μm (HW e^{-2} M).

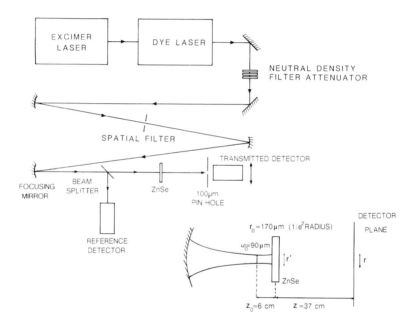

Fig. 1. Experimental set-up. Insert shows focussing geometry.

Two photodiodes were used to detect the incident energy and the transmitted fluence. The transmitted spatial profiles were measured by placing a 100 μm pinhole in front of a photodiode and traversing it across the beam, in steps of 100 μm, on a translation stage of 10 μm resolution. The profiles were measured at a distance, z, of 37 cm behind the exit plane of the sample. The pulse energy was regulated by the use of an attenuator which introduced no aberrations into the spatial profiles of the pulses. The transmitted spatial profiles were measured across the wavelength range 471-476 nm, with typical peak incident irradiances of 1 MW cm^{-2}. The temporal and spatial profiles of the incident pulses were measured at the detector plane. The theory of self-defocussing has been well developed by, and is in good agreement with, the experimental results of Weaire et al. (1979) and Guha et al. (1985). Hence this technique has been used to determine the sign and magnitude of the nonlinear refractive index. Here, we have extended the theory of Guha et al (1985) to include the carrier recombination time.

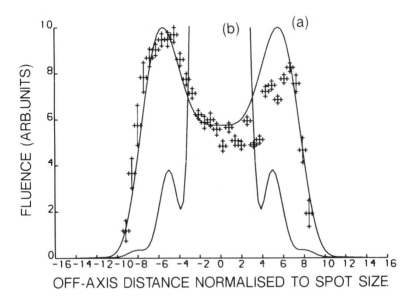

Fig. 2. Crosses are experimental points, curve (a) is fit yielding $\sigma_n = -2.5 \times 10^{-20}$ cm³. Curve (b) has $\sigma = 2.5 \times 10^{-20}$ cm³. $I_p = 1.03$ MW cm⁻² at $\lambda = 472$ nm.

Fig. 2 shows an example of a transmitted spatial profile, and its theoretical fit (curve (a)), from which the value of the refractive cross-section at that wavelength is found. The spectral dependence of σ_n found by this method is shown in Fig. 3. The values of γ and γ_R used for all the fits were 5×10^7 s⁻¹ and 5×10^9 cm³ s⁻¹ respectively. These values were obtained from theoretical fits to a temporal pump depletion experiment by Ji et al. (1989).

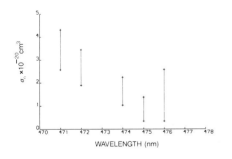

Fig. 3. Spectral dependence of σ_n.

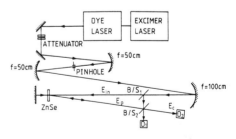

Fig. 4. DFWM experimental set-up.

DEGENERATE FOUR-WAVE MIXING

We used the technique of backward degenerate four-wave mixing in order to quantify the effective nonlinear cross-section. The experimental details and configuration are shown schematically in Fig. 4. Three optical pulses, each of 5 ns duration produced by the same laser as in the above experiment, were obtained using two beamsplitters and a mirror. The pulses were aligned so as to coincide within the sample both temporally and spatially. The angle between the pump beams and probe beam was less than one degree in the sample. The pump beams were focussed to spot sizes of 400 μm (HW e^{-2} M), and the weak probe beam was focussed onto a smaller area (100 μm HW e^{-2} M) so that it monitored a nearly uniformly irradiated region. The maximum irradiances of the pump beams were 0.5 MW cm^{-2}. The incident and phase conjugate energies were recorded by two photodiodes linked to a microcomputer.

The general theory of four-wave mixing was developed by Wherrett et al. (1983). In a nonlinear medium, the third-order polarisation for the phase-matched condition is given by:

$$P_i^{(3)}(\omega, t, z) \quad \propto \quad E_j(\omega, t, z)$$

$$\int_{-\infty}^{t} E_k^*(\omega, t', z) \, E_l(\omega, t', z) \exp\left(-\frac{t-t'}{\tau}\right) dt' \tag{1}$$

E$_j$ refers to a field polarised in the j-direction and τ is the carrier recombination time. Solving Maxwell's wave equation in the presence of such a polarisation we obtain the following expression for the conjugate irradiance:

$$I_c(t) \quad = \quad \exp\left(-\alpha D\right)(1 - \exp\left(-\alpha D\right))^2(1-R)^5 \, r \left(\frac{\sigma_{eff}}{\hbar c}\right)^2$$

$$I_0(t) \mid \quad \int_{-\infty}^{t} I_0(t') \exp\left(-\frac{t-t'}{\tau}\right) dt' \mid^2 \tag{2}$$

where we have defined

$$\sigma_{eff} \quad = \quad \sqrt{|\sigma_n|^2 + |\sigma_\alpha \frac{c}{2\omega}|^2} \tag{3}$$

By assuming that the incident (probe) irradiances are of Gaussian spatial profile with spot sizes of r$_0$ (r$_p$), the energy for the conjugate pulse is

$$\varepsilon_c \quad = \quad r(1-R)^5 \exp\left(-\alpha D\right)(1-\exp\left(-\alpha D\right))^2 \left(\frac{\sigma_{eff}}{\hbar c}\right)^2 \frac{r_p^2}{\pi^2 r_0^4(r_p^2 + 2r_0^2)} F \, \varepsilon_{in}^3 \tag{4}$$

where

$$F \quad = \quad \frac{\int_{-\infty}^{\infty} e^{-\frac{2t}{\tau}} f(t) \mid \int_{-\infty}^{t} \frac{e^{t'}}{\tau} f(t') dt' \mid^2 dt}{\left(\int_{-\infty}^{\infty} f(t) dt\right)^3} \tag{5}$$

and where f(t) is the is the time-dependence of the incident irradiance, i.e. the temporal profile.

This profile is measured experimentally, therefore F is a known constant. The experimental results confirm the cubic dependence of ε_c on ε_{in} predicted by equation (4), and taking τ as 20 ns we obtain the spectral dependence of the effective cross-section, shown in Fig. 5.

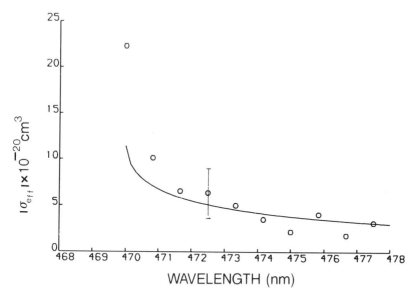

Fig. 5. Spectral dependence of σ_{eff}. Solid line is σ_n predicted by band filling theory, vertical line indicates magnitude of uncertainty.

4.   DISCUSSION

The refractive effect observed in the self-defocussing experiment can be verified to be self-defocussing by reference to Fig. 2. This shows two theoretical fits for identical parameters except for a change in the sign of the refractive cross-section. Curve (a) uses a negative value (self-defocussing), and curve (b) uses a positive value (self-focussing). If the effect was thermal in origin, it would give rise to self-focussing which is clearly not the case. This is further confirmation that the observed effect is of electronic origin.

The effective nonlinear cross-section measured in the DFWM experiment consists of a refractive and an absorptive contribution, eqn. (3); it is dominated by the refractive term for the following reason. The two photon absorption studies of Van Stryland et al. (1985), imply that $(\sigma_\alpha c/2\omega)$ is of the order of 10^{-22} cm^3, at least two orders of magnitude less than the measured effective cross-sections.

We have compared the refractive cross-sections measured by both experimental techniques with those predicted by the Drude (free carrier plasma), dynamic Burstein-Moss shift, of Poole and Garmire (1984), and band gap renormalisation, of Koch et al. (1981), models. Our results are in good agreement, in both sign and magnitude, with the three level band filling model, and hence we deduce the dynamic Burstein-Moss shift to be the dominant mechanism of these nonlinear refractive effects. There is a discrepancy of less than a factor of two between our two experimental results. According to the band filling theory, as carrier densities of $\sim 10^{18}$ cm^{-3} are approached, σ_n starts to decrease. We have estimated the peak carrier density reached in the self-defocussing experiment to be $\sim 1.1 \times 10^{17}$ cm^3. The peak irradiances used in the self-defocussing experiment were double those used in the DFWM experiment, and this may account for the discrepancy.

5.   CONCLUSION

In conclusion, we have observed self-defocussing and four-wave mixing in ZnSe at photon energies just below the band ege. We have presented theories to describe both effects, and have used them to obtain the spectra of the refractive cross-sections. Theory shows that the dynamic Burstein-Moss shift dominates this cross-section.

6.   ACKNOWLEDGEMENTS

We wish to acknowledge the financial support of the Science and Engineering Research Council (GR/E30164). Samples are prepared by N. Ross.

7.   REFERENCES

Guha S, Van Stryland E and Soileau M J 1985 Opt. Lett., **10**, 285.
Ji W, Milward J R, Kar A K, Wherrett B S and Pidgeon C R 1989 submitted for publication.
Kar A K and Wherrett B S 1986 J. Opt. Soc. Am. B, **3**, 345.
Kim, B G, Garmire E, Shibata N and Zembutsu S 1987 Appl. Phys. Lett., **51**, 475.
Koch S W, Schmitt-Rink S and Haug H 1981 Phys. Stat. Sol. (b), **106**, 135.
Landolt-Bornstein 1982 'Numerical data and functional relationships in science and technology', ed. Madelung O, **17**, b̲, 126 (Springer-Verlag, Berlin-Heidelberg-NewYork).
Lee Y H, Chavec-Pirson A, Koch S W, Park S H, Morhang J, Jeffrey A, Peyghambarian N, Banyai L, Gossard A C and Wiegmann W 1988 Phys. Rev. Lett., **57**, 2446.
Olbright G, Peyghambarian N, Gibbs H.M., MacLeod A and Van Milligen F 1984 Appl. Phys. Lett. **45**, 1031.
Peyghambarian N, Park S H, Koch S W, Jeffrey A, Potts J E and Cheng H 1988 Phys. Lett., **5**, 182.
Poole C D and Garmire E 1984 Opt. Lett. **9**, 356.
Smith S D, Mathew J G H, Taghizadeh M R, Walker A C, Wherrett B S and Hendry A 1984 Opt. Commun., **151**, 356.
Taghizadeh M R, Janossy I and Smith S D 1985 Appl. Phys. Lett. **46** 331.
Van Stryland E W, Vanherzeele H, Woodall M A, Soileau M J, Smirl A L, Guha S and Boggess T F 1985 Opt. Eng., **24**, 613.
Van Stryland E W, Wu YY, Hagan D J, Soileau M J and Mansour K 1988 J. Opt. Soc. Am. B **5**
Weaire D, Wherrett B S, Miller D A B and Smith S D 1979 Opt. Lett., **4**, 331.
Wherrett B S and Smith S D 1986 Physica Scripta **T13**, 189.
Wherrett B S, Smirl A L and Boggess T F 1983 IEEE J. Quant. Electron., **QE-19**, 680.

Inst. Phys. Conf. Ser. No 103: Part 1
Paper presented at Int. Conf. Materials for Non-linear and Electro-optics, Cambridge, 1989

Preparation of single crystal fibers

Robert S. Feigelson
Center for Materials Research,Stanford University, Stanford, CA 94305-4045

ABSTRACT:

Single crystal fibers of nonlinear, electro-optic and laser materials, both individually and bundled together in various configurations, hold promise for a variety of device applications where enhanced optical light guiding is required. A miniaturized float-zone process using a laser heat source has been found very useful for preparing single crystal fibers of a wide variety of materials including ceramics, metals, halides, etc. This method has not only been found to allow the growth of a large number of compounds in single crystal fiber form with controlled, uniform composition, but also is one of the simplest methods for preparing single crystals for property evaluation.

1. INTRODUCTION

Most electro-optic devices require single crystal components. These are most often fabricated from large bulk crystals but occasionally in the form of thin films on a suitable substrate. Single crystals in fiber form, however, provide a new possibility of combining a useful properties of an electro-optic material with the light guiding properties afforded by the fiber geometry. The benefits for single crystal optical fiber devices include higher energy densities, optical gains, and nonlinear coefficients than glass fibers, compatibility with optical glass fiber transmission media and reduced optical damage. The use of single crystal fibers either as single elements or bundled together into 2- and 3-dimensional arrays can lead to the design and fabrication of devices not possible from other types of materials. All classes of materials, including inorganic and organic compounds, semiconductors, and metals can, in principal, be produced in the form of single crystal fibers. In addition, these fiber crystals may have greater crystalline perfection than similar material in bulk form. A review of potential applications for single crystal fibers has been given by Goodman (1978) and Feigelson (1988).

2. PREPARATION OF SINGLE CRYSTAL FIBERS

Techniques for growing single crystal fibers depend to a large extent on the various thermodynamic, kinetic and physical properties of the material of interest. Various vapor, solution and melt growth techniques are available but when possible melt growth methods are most useful because they allow the most rigorous control of diameter and length. Many of these melt growth techniques are listed in Table I. Although these methods are suitable for growing fiber crystals of a variety of materials, the most versatile of these techniques is the float zone process used in conjunction with laser heating (Feigelson 1986). This method is illustrated in Fig. 1. The major advantages of this method include, 1) Steep radial and axial temperature gradients which allow rapid growth rates, enhanced grain selection and improved interface stability, 2) Crucibles or furnace refractories are not required and therefore melt contamination problems are significantly reduced, 3) Rapid heating and cooling rates are possible with laser heating as well as the ability to achieve very high temperatures, 4) Quantity of material required is minimal and, 5) Because of the ability to closely control melt composition, it is one of the best methods for growing crystals of solid solutions with excellent compositional uniformity and materials which are incongruently melting. In the latter case, the melt composition will adjust itself automatically to the proper value.

TABLE I

Melt Growth Methods for the Preparation of Single Crystal Fibers

Method	Reference
Pulling though a die	3
Edge-defined growth (EFG)	4
Float-zone (Pedestal) growth	5-7
Solidification in capillary tubes	8-10
Capillary drawing	11
Pressurized and capillary-fed growth	12
Micro Czochralski	13

Laser stability, melt volatility (due to the large surface to volume ratio) and the homogeniety and density of the starting material are problem areas important to the successful growth of fiber crystals. The magnitude of their importance, however, depends on the properties of the materials being grown. Materials with high vapor pressure should be grown by other techniques.

CO_2 lasers ($\lambda = 10.6$ μm) have been routinely used but other lasers such as Nd:YAG (Bagdasarov) can also be effective. The ultimate choice of laser wavelength and power capability depends on the materials to be grown and certain other parameters such as fiber diameter. The power necessary to maintain a stable molten zone decreases with diameter, as shown in Fig. 2. With the use of beam shaping optics, it is possible to convert a solid laser beam into an annular ring shape which can then be tightly focussed onto a cylindrical source rod, giving circularly symmetric heating (Fejer et al 1982). Beam widths of 25 μm can be achieved and are useful for the growth of very small diameter fibers. Starting material can be fabricated from single crystals or polycrystalline solids prepared by various techniques (for ceramics, cold pressing - sintering and hot pressing). The requirement for seed crystals are minimal because grain selection is very effective in small diameter samples. Oriented crystals are possible through the use of suitably oriented seed crystals.

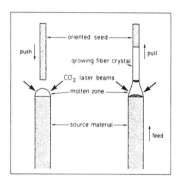

Fig. 1. Schematic of laser heated float zone (pedestal) growth process.

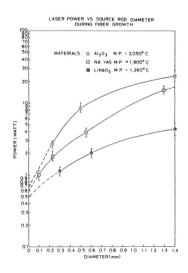

Fig. 2. Relationship between power required to melt source material of Al_2O_3, Nd:YAG and $LiNbO_3$ as a function of source rod diameter.

3. CRYSTAL QUALITY

For optical devices, low internal scattering and absorption losses, and constant diameters are required. Crystal quality depends on a number of factors, including the stoichiometry and purity of the starting material, the growth rate, temperature gradients, temperature stability, sample size, melt convection, etc. In the miniaturized laser-heated float zone process, the steep temperature gradients help promote interface stability and fairly rapid growth rates (mm/min) can be used for pure melts. Contamination from crucible materials or furnace refractories is not a factor so potential absorption losses and scattering by second phase particles from this source is eliminated.

Chemical inhomogeneties can give rise to birefringence variations in optical materials and, hence wavefront distortion. The float zone process tends to minimize these inhomogeneties because the melt composition is closely controlled by the source rod composition. Since material of constant composition is continuously being fed into the melt during growth, the growing crystal must eventually reach this composition. In fibers, this steady state composition is reached within a few fiber diameters. Some radial chemical inhomogeneties can arise when dopants are used due to the possibly different segregation coefficients at facet planes and edges on the growth interface. This can be controlled, in principal, by altering interface shape by changing processing parameters.

Another source of optical losses and poor electric-optic properties can come from lattice defects such as grain boundaries, twins, dislocations, multiple domains. Many such defects can be substantially lower in small diameter fibers due to a number of factors. For example, the narrow diameter of the crystals permits line defects to grow out if they are not aligned exactly along the fiber axis. Twins and dislocations present in the seed crystal or develop during the initial seeding process usually grow out after a short time. Tsivinsky (1968) and Inoue and Komatsu (1979) have shown that dislocation densities will decrease dramatically as the crystal diameter is reduced below some value, which depends on the axial temperature gradients during growth and certain material parameters; the Burgers vector, critical resolved shear stress, the shear modules and the thermal expansion coefficient of the material.

For the above reasons, small diameter electric-optic crystals should have greater perfection than their larger counterparts. Therefore, it should be possible to more closely measure the intrinsic properties of these materials rather than properties dependent on defects. This would allow a better assessment of their properties for electric-optic applications.

Both short and long range diameter variations have been observed in fibers grown by various techniques and their magnitude and periodicity are strongly related to material and growth system parameters which are reflected in changes in melt volume and shape. Diameter control during growth is necessary to minimize optical losses. Laser and mechanical stability in the growth station and thermal symmetry are important, but when these parameters are fixed, different materials will exhibit greater or lesser diameter variations, depending on their specific properties. There are a number of possible schemes for controlling single crystal fiber diameter, an effective one having been developed by Fejer (1986) using a laser scattering technique (Fig. 3) to detect diameter changes and making adjustments to feed rates as needed. This method also permits the introduction of periodic (modulated) diameter variations of controlled spacing (Fig. 4).

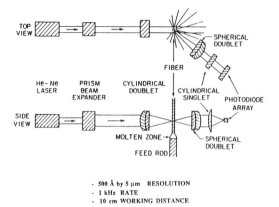

Fig. 3. Schematic of automatic diameter control system for
fiber growth (Fejer 1986).

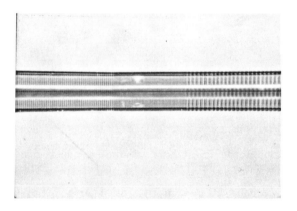

Fig. 4. Modulated fiber diameter produced with automatic
diameter control system (Fejer 1986).

4.   FIBER CLADDING

In addition to uniform diameter and good optical quality of the as-grown single crystal fiber,
light guiding is enhanced by cladding the fiber with a low index coating similar to glass fiber
elements. Work on developing fiber cladding techniques for single crystal fibers has been
underway for a number of years. Three approaches have been studied at Stanford University
to date (Digonnet,Sudo), including the extrusion of a thin index-matched glass coating on
Nd:YAG laser fibers (Digonnet) and the creation of a reduced refractive index layer in the
surface of single crystal MgO doped $LiNbO_3$ fibers by diffusion or an ion exchange process.

Nd:YAG fibers clad with a thin film of index-matched glass produced cw fiber lasers with
slope efficiencies (ratio of laser output power to absorbed pump power in excess of

threshold) of about 20% in clad devices in which both the signal and the pump were guided. The fiber propagation loss factors deduced from laser characterization measurements were several times smaller than the best loss figures previously measured in unclad fibers.

Laser Q-switching was also successfully demonstrated in clad and unclad Nd:YAG fibers. Pulses as short as 8 ns were generated in a clad, guiding fiber laser. These results constitute the first demonstration of Q-switched operation in a fiber laser.

Another cladding approach being investigated by Sudo et al. involves doping the outside portion of MgO:LiNbO$_3$ fibers with a suitable element so as to locally reduce its refractive index (i.e. lower than that of the fiber material by 0.003 to 0.10). Doping can be achieved in a number of ways, including diffusion, ion-exchange and ion-implantation. These techniques present several advantages over the glass coating technique used for Nd:YAG. First, they make possible the introduction of controlled concentrations of impurities in the fiber, and therefore the fabrication of a cladded region having well defined index difference and depth. Second, they provide a means of reducing the size of the core region in which the optical wave is to be confined, by as much as a factor of two for concentration independent diffusion processes. For the case of Mg diffusion, a 30 μm diameter fiber will have a 15 μm core diameter after diffusion. This removes the necessity of growing and handling very small diameter fibers. Third, the diffusion process itself smooths out diameter variations with spatial periods smaller than the diffusion depth, i.e. smaller than 10-30 μm, which will further reduce surface scattering loss. Fourth, the impurities are introduced in the cladded region (where the electric field of the optical wave is relatively small), unlike in optical circuits designed for surface waveguide applications where they are present in the core region of the guide where the field is relatively strong. Any deleterious effects that these impurities may have on the crystal quality and/or properties therefore will have considerably reduced influence on the optical wave. Finally, this process eliminates the difficulties encountered when interfacing two different materials.

Four dopants have been studied as potential candidates for cladding lithium niobate: magnesium, tantalum and zinc by diffusion, and protons (by ion-exchange). Recently, studies on Zn doped LiNbO$_3$ have yielded very promising results (Young).

5. METASTABLE PHASE FORMATION

Due to the high growth velocities typically used in laser-heated float zone fiber growth and the steep temperature gradients present (> 1000° C/cm), metastable high temperature phases can persist to room temperature [ie., BaTiO$_3$, ScTaO$_4$ (Feigelson 1986,Elwell 1985)] with good structural and optical integrity. This may permit the preparation and characterization of useful new electric-optic materials which could not be produced in single crystal form before by conventional techniques.

6. GROWTH OF ELECTRO-OPTIC CRYSTALS

6.1 LiNbO$_3$

Luh et al. (1986) showed that LiNbO$_3$ fibers grown along the c-axis by the laser heated float zone method had a single domain (ferroelectric) structure when the diameter was less than 700 μm and a bi-domain structure when grown along the a-axis. These domain configurations were attributed to the thermoelectric effect which generates an electric poling potential during growth due to the large temperature gradients in the system. Luh (1988) also showed that when the LiNbO$_3$ source rod was melted with a two beam laser system and rotated, a periodic domain structure resulted due to the asymmetric radial temperature distribution (Fig. 5). This work was expanded upon by Magel et al.(1989). They have produced highly periodic domain structures in LiNbO$_3$ using various means of modulating the microscopic growth rate during pedestal growth (see Fig. 6). They were able to create structures with domain reversals every 1 μm in both *a*-axis and *c*-axis oriented crystals, and in both Mg-doped and undoped material. This material has been applied to the generation of microwatts of blue light at wavelengths as short as 407 nm by room-temperature quasi-phasematched frequency doubling of a cw infrared dye laser (Magel et al).

Fig. 5. Periodic domains induced by crystal rotation
in an asymmetric thermal gradient (Luh 1988).

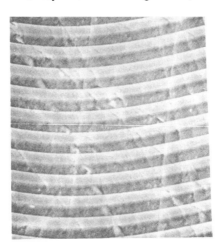

Fig. 6. SEM photograph of periodically alternating domains with
thickness under 2 μm as revealed by polishing and
etching the *y* face of an *a*-axis pedestal-grown crystal.
The curvature of the domains is due to the curved shape
of the growth interface (Magel et al 1989).

6.2   BaTiO₃

$BaTiO_3$ undergoes a high temperature phase transformation from hexagonal to cubic structure. When small diameter crystals are grown in steep temperature gradients, the hexagonal phase can be produced in the form of a high optical quality single crystal fiber. Saifi et al. (1986) showed that when a $SrTiO_3$ cubic seed crystal is used to initiate growth, the $BaTiO_3$ will crystallize directly into the cubic phase, by-passing the hexagonal phase. This may be due to the expitaxial deposition of $BaTiO_3$ on the $SrTiO_3$ seed crystal.

6.3 $Sr_xBa_{1-x}Nb_2O_6$

Strontium barium niobate ($Sr_x Ba_{1-x}Nb_2O_6$) is a promising photorefractive material for applications such as data storage, optical correlation and convolution and associative memory. Bulk crystals of high quality have been difficult to grow by the Czochralski method and while significant progress has been made in recent years (Neurgaonkar), this material is still not available commercially. An alternate strategy using SBN single crystal fibers and fiber arrays has been pursued by Hesselink and Redfield (1988). This approach is attractive because of the greater ease of growing SBN fibers compared with bulk crystals, the prospect of higher crystalline perfection, and the ability to make multiple fiber devices. These photorefractive fibers are an attractive recording medium because they permit waveguiding over long propagation lengths and can be bundled together to give large apertures. Fiber arrays can be made in substantially larger volumes than is possible with the bulk SBN crystals currently available and these arrays can be in various geometric patterns which can provide an added degree of freedom for image-processing applications such as pattern recognition in large scale recording medium. The fiber matrix array allows the photorefractive properties of the individual fibers to be independently varied with voltage in temperature.

Strontium barium niobate fibers were first grown by Kway and Feigelson in 1984 using the laser heated float zone process and single crystal source material of composition $Sr_{.46}Ba_{.54}Nb_2O_6$ and $Sr_{.75}Ba_{.25}Nb_2O_6$. Two c-axis fibers were grown at 0.62 and 0.56 mm/min respectively in an ambient air atmosphere. Both crystals were colorless as grown and did not contain striations. The first fiber had dimensions of 600 μm diameter and 55 mm in length and the second was 1700 μm diameter and 30 mm long (Fig. 7). These fibers did not exhibit the pronounced circumferential faceting usually observed in Czochralski grown boules. Hesselink and Redfeld (Hesselink) grew Ce doped $Sr_{.60}Ba_{.40}Nb_2O_6$ fibers in diameters ranging from 160 mm to 2mm and 4-8 mm in length using this same method. They found that the angular sensitivity for holographic recording appears to be considerably improved using fiber crystals.

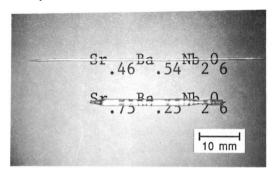

Fig. 7. Two fibers of SBN of 600 μm and 1700 μm diameter
(Kway, Feigelson 1987-8).

Work on the growth of SBN fibers was reported recently by Yamamoto and Bhalla (1989). They used the same method described above and had similar results.

7.   CONCLUSIONS

Exciting possibilities exist both for the study of the properties of electro-optic materials in the form of high quality single crystal fibers and for the development of electro-optic fiber devices with improved properties and in new device configurations. The miniaturized float zone process using laser heating is a useful technique for preparing such fibers for characterization and for use in devices, but its suitability depends on the materials under investigation and the fiber requirements.

8.   REFERENCES

Bagdasarov Kh S, Dyachenko V ., Kevorkov A . and Kholov A 1986 in: Growth of Crystals 13, Ed. E. I. Givargizov, Consultants Bureau, N. Y. p.364.

Bridges T J, Hasiak J S and Strnad A R 1980 Opt. Letters **5** 1985.

Burrus C A and Stone J 1975 Appl. Phys. Letters **26** 318.

Digonnet M J F, Gaeta C, O'Meara D and Shaw H J 1987 IEEE J. Lightwave Tech. **5** 642.

Elwell D, Kway W L and Feigelson R S (1985) J. Crystal Growth **71** 237.

Fejer M M, Byer R L, Feigelson R S and Kway W 1982 Proc. SPIE, Advances in Infrared Fibers II **320** 50.

Fejer M M 1986 Ph.D. Thesis, Stanford Univ.

Feigelson R S, Growth of Fiber Crystals, in: Crystal Growth of Electronic Materials, Ed. E. Kaldis (North-Holland, Amsterdam, 1985) p. 127.

Feigelson R S 1986 J. Crystal Growth **79** 669.

Feigelson R S 1988 Materials Sci. & Eng. **B1** 67.

Goodman C H 1978 Solid State Electron Devices **2** 129.

Haggerty J S, Production of Fibers by a Floating zone Fiber Drawing Technique, Final Report NASA-CR 120948 (May 1972).

Hesselink L and Redfield S 1988 Optics Letters **13** 877.

Inoue T and Komatsu H 1979 Kristell Tech. **14** 1511.

Kway W L and Feigelson R S July 1 1985-April 30, 1986 Center for Materials Research 24th Annual Report on Materials Research at Stanford University -CRM-85-2.

LaBelle H E Jr and Mlavsky A I 1922 Mater. Res. Bull. **6** 184.

Luh Y-S 1988 Ph.D. Thesis, Stanford Univ.

Luh Y-S, Feigelson R S, Fejer M M and Byer R L 1986 J. Crystal Growth **78** 135.

Magel G A, Fejer M M and Byer R L April 24-28, 1989 Paper ThQ3, Conference on Laser and Electro-Optics, Baltimore, MD .

Mimura Y, Okamura Y, Komazawa Y and Ota C 1980 Japan J. Appl. Phys. **19** L269.

Neurgaonkar R R, Cory W K and Oliver J R 1983 Ferroelectrics **51** 3.

Saifi M, Dubois B, Vogel C M and Thiel F A 1986 J. Mater. Res. 1 452.

Stevenson J L and Dyott R B 1974 Electron. Letters **10** 449.

Sudo S, Cordova-Ploza A, Byer R L and Shaw H J 1987 Opt. Lett. **12** 938.

Tripp M 1986 private communication.

Tsivinsky S V 1968 Fiz Metallov Metalloved. **25** 1013.

von Gomperz E 1922 Z. Physik **8** 184.

Weber H P, Liao PF, Tofield B C and Bridenbaugh P M 1975 Appl. Phys. Letters **26** 692.

Yamamoto J K and Bhalla A S 1989 Mat. Res. Bull. **24** 761.

Young W-M, Shaw H J, Fejer M M, Digonnet M J F and Feigelson R S 1989 private communication.

Inst. Phys. Conf. Ser. No 103: Part 1
Paper presented at Int. Conf. Materials for Non-linear and Electro-optics, Cambridge, 1989

85

Laser densification of sol-gel silica glass

D J Shaw, A J Berry and T A King

Physics Department, Schuster Laboratory. University of Manchester, Manchester M13 9PL.

ABSTRACT: Development of densification techniques for sol–gel matrices using laser radiation is presented. A carbon dioxide laser has been shown to be effective in producing full densification of thin surface layers of silica sol-gel. Microhardness as a characterisation method of the lower energy densification threshold and the upper (bloating) threshold and refractive index as a monitor of densification have been developed. Energies to effect full densification were measured as 6.3 ± 0.3 and 4.3 ± 0.2Jcm^{-2} of surface irradiated.

1. INTRODUCTION

The generally accepted structure for partially dense gel–silica is a microporous solid based on an SiO_2 skeleton (Klein 1988). Most of the pores are open and are, on average, 1.2nm diameter. Water molecules are hydrogen bonded to oxygen atoms on the pore surface as well as OH^- ions. Gallo et al (1984, 1987) describe the physical and chemical processes occurring during densification as

1. Physical desorption of hydrogen bonded water below 200$°$C

2. Decomposition of residual organic compounds over 300$°$C – 500$°$C

3. Pore collapse over 500$°$C – 1150$°$C.

OH^- ions are steadily eliminated by condensation polymerisation as below:

$$Si - OH + HO - Si \longrightarrow Si - O - Si + H_2O$$

This process increases the surface energy of the pores. OH^- ions are largely eliminated at temperatures $>1100\,°C$. The mechanism of pore collapse, known as viscous sintering, occurs because the gel-silica tends to a minimum energy state. In this instance a minimum energy state corresponds to minimum surface energy, hence the pores will tend to collapse. We find that a viscous relaxation of the pores takes place which is activated by heating the gel-silica. By about $1150\,°C$ densification is complete, giving a gel silica with a density equal to that of conventional vitreous silica, ie $2.21\,gcm^{-3}$. The energy required to effect full densification may be estimated by determining the energy required to raise the temperature of a surface layer of gel silica to $1150\,°C$. This energy H is

$$H = mc\Delta\theta \qquad\qquad (1)$$

where m is the mass of the surface layer, c is the specific heat, and $\Delta\theta$ is the temperature change.

It is convenient to work with the energy required per square centimetre of surface irradiated. Since the laser radiation is significantly absorbed by silica and is not absorbed in the pores we are justified to use the density of bulk silica, ie $2.21\,gcm^{-3}$, in calculating the mass. Taking the $1/e$ penetration depth of $10.6\mu m$ radiation in gel-silica as $40\mu m$ every cm^2 of surface irradiated is equivalent to irradiating a volume of $1 \times 0.004 = 4 \times 10^{-3}cm^3$. Hence every cm^2 irradiated is equivalent to irradiating a mass $m = 2.21 \times 4 \times 10^{-3} = 8.84 \times 10^{-3}g$. For $\Delta\theta = 1150 - 20 = 1130\,°C$ and $c = 0.753$ $J/g\,°C$ the energy required is $7.4\,Jcm^{-2}$.

This research investigates the feasibility of laser induced densification, the energies required and the properties of laser densified gel-silica. Techniques are being developed for the laser writing of refractive index profiles, in particular waveguides and densification of large surface areas.

2.   EXPERIMENTAL PROCEDURE

Sol-gel prepared polymeric silica (Hench et al 1988) in the form of 1 inch diameter discs was supplied by GELTECH Inc. The samples had been pre-densified to $800\,°C$ by conventional thermal methods. This means that the samples provided have a density of $1.35\,gcm^{-3}$ (with ~ 28% of the volume being pores), a microhardness of $190\,kg\,mm^{-2}$ and a refractive index of 1.42. Fully dense silica glass has a density of $2.21\,gcm^{-3}$ a microhardness of $680\,kg$ mm^{-2} and a refractive index of 1.46. The laser written tracks were produced by mounting the gel-silica disc on a pendulum bob which is swung through the focused carbon dioxide laser beam once and then stopped. As the pendulum swings through the laser beam it interrupts a second helium-neon laser beam incident normally on a photodiode which is connected to an oscilloscope, thus the speed of the pendulum, and hence the incident energy can be calculated. The carbon-dioxide laser (Ferranti, model CM2000) produced up to 17 watts with a focused beam radius of up to 380µm. In general a densified/damaged track is written across the disc. Densified material was characterised by measuring the Vicker's microhardness with a Riechart 2683 model microhardness apparatus. Loads of 24g and in some cases 32g and 40g for harder material were used. None of these load values were found to be such as to exceed the stress factor when applied to appropriate regions. Refractive index methods based on a reflectivity technique to characterise laser written tracks are being developed.

3.   RESULTS

Figure 1 shows some photographs of laser treated tracks taken with an Olympus AH2 photomicroscope. Figure 1a is a laser treated line with an energy of 2.34Jcm^{-2}, ie below threshold, here the beam has just 'cleaned up' the surface. Figures 1b and 1c show lines exposed to 6.67 and 5.98Jcm^{-2}. Both these photographs were taken under dark field photography; this method illuminates scattering centres brightly while smooth areas appear dark. Figure 1d shows a track produced by an energy density of 11.53Jcm^{-2}; the energy here is above the damage threshold, the gel-silica has been exposed to an energy above the damage level and bubbles of released gases have formed. The gel is said to bloat. Bloating occurs when gases and/or water vapour released during the densification process cannot escape because the pores have already collapsed. Thus in Figure 1b we see a bright bloated line whilst Figure 1c is mainly dark indicating a smooth and densified track. Thus we can deduce that the bloating, or upper densification threshold, is between 6.67 and 5.98Jcm^{-2} giving a mean upper densification threshold of 6.32 ± 0.34Jcm^{-2}. Figure 1e is a photograph of a densified track under interference photography.

Figures 2a and 2b are microhardness profiles of various laser treated tracks with different incident energies. The microhardness profiles are gaussian because the transverse profile of the CO_2 laser beam is gaussian. Fully dense silica has a microhardness of 680kg mm^{-2}. The three curves in figure 2a have peaks within errors of this value whereas the two curves of figure 2b are not within this value. We deduce that the lower densification threshold is between 4.45 and 4.06Jcm^{-2}, with a mean value for the lower densification threshold of 4.26 ± 0.20Jcm^{-2}.

4.   DISCUSSION AND CONCLUSIONS

The upper and lower densification thresholds have been measured as 6.32 ± 0.34Jcm^{-2} and 4.26 ± 0.20 Jcm^{-2} respectively. These energies using equation 1 correspond to raising the exposed gel-silica to temperatures of 950°C and 640°C. The values 6.32Jcm^{-2} and 4.26Jcm^{-2} are average beam intensities; since the laser beam has a gaussian profile the peak intensity is approximately twice (or exactly 2.066 x) the measured average intensity, thus the peak intensity thresholds are 13.06Jcm^{-2} and 8.80Jcm^{-2}, which correspond to raising the temperature of the gel-silica to 1962°C and 1322°C. This lower threshold temperature is in reasonable agreement with the minimum temperature for densification of 1150°C. The lower energy threshold has been calculated as 7.42Jcm^{-2} as an average value; the peak intensity for the measured lower densification is 8.80Jcm^{-2}, again this in good agreement with the calculated value.

Figure 1: Densified/Damaged Tracks

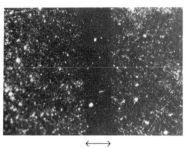

Figure 1a   ←——→ 200μm

Energy = 2.34Jcm^{-2}

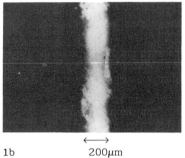

1b   ←——→ 200μm

Energy = 6.67Jcm^{-2}

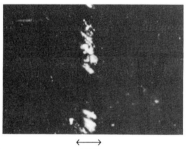

1c   ←——→ 200μm

Energy = 5.98Jcm^{-2}

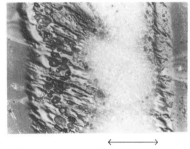

1d   ←————→ 200μm

Energy = 11.53Jcm^{-2}

1e   200μm

Energy = 4.72Jcm^{-2}

Figure 2: Microhardness profiles of laser written tracks with different incident energies

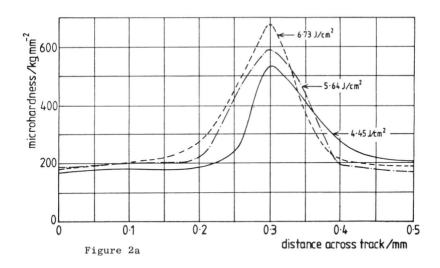

Figure 2a

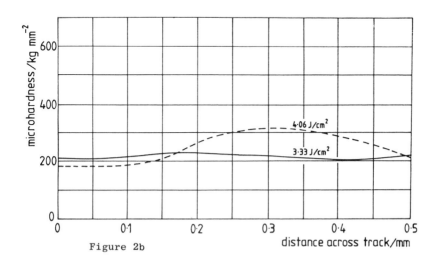

Figure 2b

REFERENCES

1. Gallo T A, Brinker C J, Klein L C, Scherer G W, 1984 "The Role of Water in Densification of Gels", "Better Ceramics Through Chemistry" Ed, C J Brinker, D E Clark and D R Ulrich, Amsterdam (North-Holland), pp85-90

2. Gallo T A, 1987 "Densification of Gel-Derived Silica", PhD Thesis, Rutgers State University of New Jersey, pp4-5

3. Hench L L, Wang S H and Nogues J L, 1988 Proc SPIE 878

4. Klein L C (Ed), 1988 "Sol-Gel Technology for Thin Fibres, Preforms, Electronics and Speciality Shapes", New Jersey (Noyes Publications)

ACKNOWLEDGEMENTS

This study forms part of a wider programme of research in collaboration with the Advanced Materials Research Center, University of Florida and GELTECH Inc. We have enjoyed extensive discussions with Professor Larry Hench of the ARMC and we thank Mr Bill Moreshead of GELTECH Inc for discussion and provision of samples. This research has been sponsored by the Air Force Office of Scientific Research (AFSC) under contract F49620-88-C-0010. The United States Government is authorized to reproduce and distribute reprints for governmental purposes notwithstanding any copyright notation hereon.

Experimental determination of thermal nonlinearities of semiconductor-doped glasses

S. De Nicola[1], P. Mormile[1], G. Pierattini[1], G. Abbate[2], U. Bernini[2], P. Maddalena[2], G. Assanto[3]

[1]Istituto di Cibernetica del CNR - Via Toiano, 6, 80072 Arco Felice (Napoli), Italy

[2]Dipartimento di Scienze Fisiche - Pad. 20, Mostra d'Oltremare, 80125 Napoli, Italy

[3]Dipartimento di Ingegneria Elettronica - Viale delle Scienze, 90128 Palermo

ABSTRACT

An esperimental method for the determination of thermal nonlinearities in semiconductor-doped glasses induced by c.w. laser radiation is presented. An estimation of the intensity-dependent refractive index is given, at steady-state conditions, which is in good agreement with experimental data reported in literature for similar materials.

Semiconductor doped glasses have received a great deal of attention in recent years. In particular, nonlinear optical properties of glasses doped with Cd S_x Se$_{1-x}$, available commercially as color filter glasses, have been extensively studied. Estimation of the steady state value of the intensity dependent refractive index n_2 of Schott GG-495 glass filter has been made by Patela et al. [1].

The value $n_2 = 8 \times 10^{-11}$ m^2/W is four orders of magnitude greater than the one obtained with short laser pulses by Jain and Lind[2]. The discrepancy was ascribed to a thermal contribution to the nonlinear refractive index due to the use of cw laser radiation. In order to investigate the thermal nonlinearity of doped glasses we performed measurements of the temperature dependence of the refractive index of Corning sharp cut filter Cs 3-69, 4.6 mm thick. The experimental method used (see fig. 1) is based on a simple interferometric technique[3], which allows the determination of the temperature dependence of the refractive index of transparent liquids and solids [4]. An He-Ne laser beam($I_p = 1$ mW) impinges normally on the glass sample.

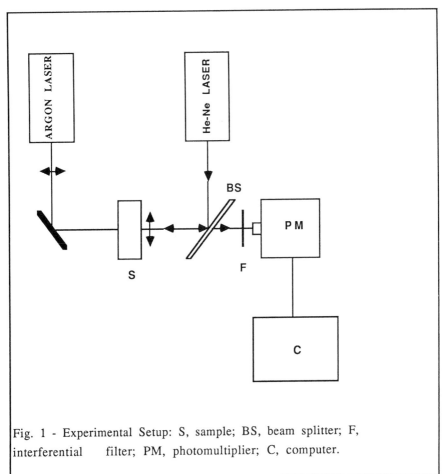

Fig. 1 - Experimental Setup: S, sample; BS, beam splitter; F, interferential   filter; PM, photomultiplier; C, computer.

The light multiply reflected from the parallel surface of the filter forms an interference pattern which is detected by a photomultiplier connected to the acquisition port of the computer. Because of the low reflectivity of the filter only the effect of the reflection from the front and back faces of the sample is observable. The intensity fringe pattern I_r is

$$\frac{I_r}{I_p} = \rho + \rho\tau^2 + 2\rho\tau\, \cos\varphi \tag{1}$$

In eq. (1) ρ and τ are the reflection and transmission coefficient and φ the phase difference between the two interfering beams.

Variations of the optical path length inside the filter are caused by the absorption of the light of a cw Ar+ laser with a focal spot radius w = 150 μm at $1/e^2$ intensity and a power of 200 mW. The He-Ne beam probes the central part of the region illuminated by the pump beam. The laser is switched on at t = 0. During the irradiation the temperature of the sample changes and the amplitude of the signal detected by the photomultiplier passes through an alternate series of maxima and minima (see fig. 2).

The phase change in eq. 1 near the beam axis is given by

$$\varphi = \varphi_0 + \left(\frac{4\pi}{\lambda}\right) d \left[n_0\beta + \left(\frac{\delta n}{\delta T}\right)_p \right] <T> \tag{2}$$

In eq. 2 the second term is the intensity dependent correction to the phase $\varphi_0 = 4\pi\, n_0 d/\lambda$ due to the temperature dependence of the refractive index n of the filter which gives rise to the detected fringe pattern, d is the glass thickness, λ = 632.8 nm the probe wavelength, n_0 = 1.507, β = 43 x 10^{-7} °K^{-1} the thermal expansion coefficient and

$<T> = \dfrac{1}{d} \int_0^d T dz$ the mean temperature along the z-direction normal to

the sample faces.   From the heat equation we calculate the temperature rise of the filter averaged along the z direction with conditions T = 0 at r = a, a=3 cm being the boundary of the medium

$$<T> = \dfrac{2P_0 a^2}{\pi k d w^2} \, [\, 1 - e^{(-\alpha d)}] \sum_{p=1}^{\infty} \dfrac{A_p}{\lambda_p^2} [\, 1 - e^{\left(\dfrac{D\lambda_p^2 t}{a^2}\right)}\,] \qquad (3)$$

with A_p given by

$$A_p = \dfrac{2}{a^2 J_1^2(\lambda_p)} \int_0^a e^{\left(-\dfrac{2\xi^2}{w^2}\right)} \, J_0 \left(\dfrac{\lambda_p \xi}{a}\right) \xi \, d\xi$$

In eq. 3 α = 10cm^{-1} is the absorption coefficient, P_o = 200 mW the incident power, k = 1W/mK the thermal conductivity, D = 5x10^{-7} m^2/s the thermal diffusivity, λ_p the zeros of Bessel function $J_0(x)$, J_1 the Bessel function of first order .   When t >> a2/4D the thermal equilibrium is attained and eq. 3 gives the steady state value T_e of the temperature(5,6) proportional to the incident pump power

$$T_e = \dfrac{P_0}{4\pi k d} [\, 1 - e^{(-\alpha d)}] \, \ln \left(\dfrac{2\gamma a^2}{w^2}\right) \qquad (4)$$

where $\gamma = 1.781$

As the temperature of the sample approaches its equilibrium value the distance between the maxima increases in time. Theoretical calculations based on eq. 2 and 3 are shown in fig. 3. The fitting with experimental data gives $(\delta n / \delta T)_P$ = 1.502 x 10^{-5}K^{-1}. For an increment of the temperature field T_e we have an index refraction variation

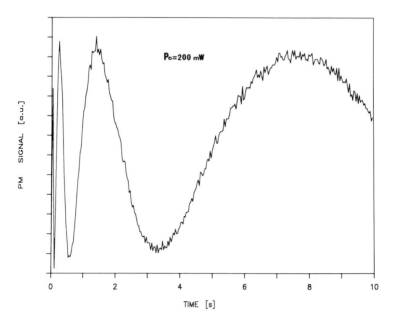

Fig. 2 - Experimental records of the photomultiplier signal for P_0 = 200mW

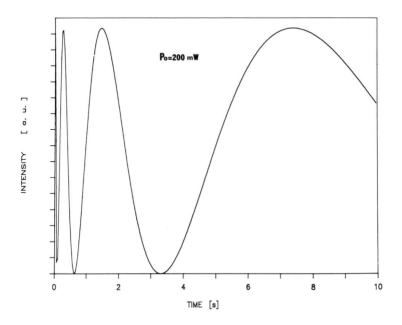

Fig. 3 - Calculated intensity fringe pattern vs. time

$\Delta n = (\delta n / \delta T)_P \cdot T_e = 5.70 \times 10^{-4}$. The nonlinear refractive index n_2 calculated as $\Delta n / I_0$ is equal to $1.1 \times 10^{-10} \, \text{m}^2/\text{W}$.

Acknowledgements

We thank Mr. S. Piscitelli and Mr. A. Finizio for their technical support

References

1) S. Patela, H. Jerominek, C. Delisle, R. Tremblay, J. Appl. Phys. <u>60</u>, 1951 (1986)

2) R. K. Jain, R.C. Lind, J. Opt. Soc. Am.<u>73</u>, 647 (1983)

3) S. De Nicola, P. Mormile, G. Pierattini, G. Abbate, U. Bernini, P. Maddalena, Opt. Commun. <u>70</u>, 502 (1989)

4) G. Abbate, A. Attanasio, U. Bernini, E. Ragozzino, F. Somma, J. Phys. D. Appl. Phys. <u>9</u>, 1945 (1976)

5) J. Stone, J. Opt. Soc. Am. <u>62</u>, 327 (1972)

6) J. Gordon, R.C. Leite, R.S. Moore, S.P.S. Porto, J. Whinnery, J. Appl. Phys. <u>36</u>, 3 (1965)

Inst. Phys. Conf. Ser. No 103: Part 2
Paper presented at Int. Conf. Materials for Non-linear and Electro-optics, Cambridge, 1989

Design and synthesis of organic nonlinear optical materials

Michael G Hutchings[*], Paul F Gordon, and John O Morley

ICI Colours and Fine Chemicals, Fine Chemicals Research Centre, Blackley, Manchester M9 3DA, UK

Abstract: The status of the design and synthesis of organic nonlinear optical materials is reviewed. Results of theoretical calculations can be used to predict the behaviour of speculative molecules. The methods used to enforce noncentrosymmetry, important for second order nonlinearity, are emphasised. Examples include ICI's crystalline 2-pyrazolines, and polyene systems studied by the Langmuir–Blodgett technique, as well as some recent literature results.

1. INTRODUCTION

A reflection of the increasing interest in, and technical importance of, organic materials for nonlinear optics applications is given graphically in Fig.1. This shows the number of patents on organic nonlinear optical (NLO) materials filed per year. While various interpretations can be put on this plot, it is at least clear that many commercial concerns believe that new organic NLO materials have a technical future sufficient to warrant increasing research expenditure, and that if we wish to compete in this area it will be necessary to continue to be innovative. Central to this endeavour is the design and synthesis of new materials.

The search for new, better NLO materials is influenced by, and depends on, various factors. A new target species or system must obviously have properties consistent with the intended technical application (eg. frequency doubling, electrooptic modulation, etc), and it must also meet materials fabrication requirements, as well as performance specifications. The intention of this paper is to review some recent progress in the area of design and synthesis of second order organic NLO materials; third order NLO materials will not be discussed. The feature which renders the former materials unique in the whole field of applied organic chemistry is their crucial requirement for a noncentrosymmetric (acentric) arrangement of the active chromogens. For convenience, the discussion in this paper will be organised roughly on the basis of this property. Examples will be taken both from ICI results and from the recent open literature.

2. THEORETICAL APPROACHES

The synthesis programme in ICI has proceeded simultaneously with, and aided by, theoretical calculations of the NLO properties of organic molecules. The theoretical method used is based on semiempirical molecular orbital theory, using a modification of CNDO programs which include configuration interaction treatments. The resultant CNDOVSB program has been

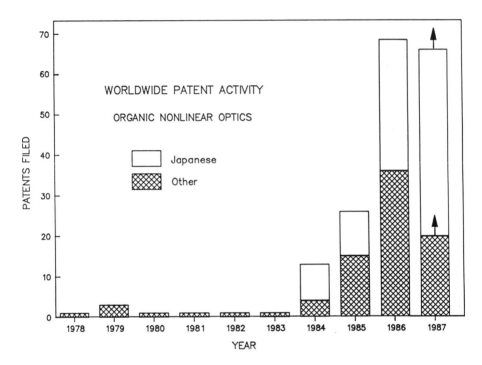

Fig.1 Worldwide organic nonlinear optics patents, up to and including
       Derwent abstract week 8910. Patents filed 1987 are still appearing.
       ----

parametrised specifically to calculate first hyperpolarisabilities (β),
both at zero field strengths and at user-stipulated energies. The program
also gives electronic absorption wavelengths and oscillator strengths, as
well as ground and excited state dipole moments (Docherty et al (1985)).

An early application was to substituent effects in benzenoid systems.
Calculations predicted that the pi-acceptor nitroso group has a
strengthening effect on β even greater than nitro (Morley et al 1987a).
This was subsequently verified experimentally. Rather more useful in
practice is the prediction, again verified experimentally, that alkylthio
is almost as good a pi-donor as the amino group with respect to β but does
not have such a bathochromic effect on the absorption maximum (Morley et al
1987b). This is of obvious relevance to materials to be used for frequency
doubling where transparency is paramount. The CNDOVSB program has been
used routinely in ICI to pre-screen the β-values of hundreds of speculative
molecules dreamed up by synthetic and theoretical chemists. Further study
of the electronic properties of both ground and excited states has then led
to better understanding of how substituent variation is likely to influence
properties of interest. Examples are discussed in context below.

3. CRYSTALLINE PHASE - 2-PYRAZOLINES

ICI's interest in 2-pyrazolines (1) as speculative NLO species was
stimulated by a report of the difference in ground and excited state dipole
moments for some simple derivatives (Gusten et al 1977). It was decided to

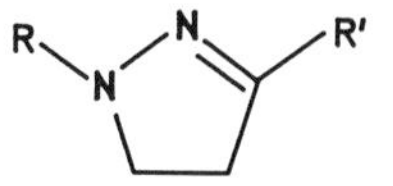

(1)

(3) R = R' = Ph; β = 18 x 10^{-30} esu

(4) R = 4—MeOPh, R' = 4—PhNO$_2$; β = 60 x 10^{-30} esu

(5) R = 4—O$_2$NPh, R' = 4—PhOMe; β = 14 x 10^{-30} esu

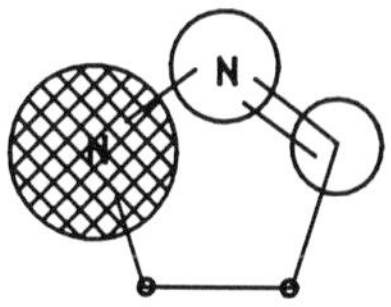

(2)

synthesise examples with the intention of discovering an NLO active crystal. Simultaneously, a computational study of the chromogen was undertaken. Application of the CNDOVSB program to the calculation of the β-values of unsubstituted 2-pyrazoline (1; R = R' = H) indicated a shift of electron density on electronic excitation as shown in (2) (Morley et al 1989). The hatched area represents electron donation, while the open circles imply build up of negative charge. Qualitatively, the chromophore is characterised by transfer of electron charge from nitrogen-1 to its adjacent atom, nitrogen-2, and to a lesser extent to the conjugated carbon-3. Simple perturbation arguments suggested that a pi-donor group at position 1, and a pi-acceptor at the 3-position, should lead to an increase in β. In fact, calculated values for (3), (4) and (5) support this conclusion. It was therefore clear that synthesis should focus on derivatives substituted in this way.

In the event, many 2-pyrazolines were synthesised and their second harmonic generation (SHG) properties studied by the Kurtz and Perry (1968) powder method. Not unexpectedly, the SHG activity of many samples was found to be zero, due to head-to-tail dipole orientation in the crystal. However, two new molecules (6; DCNP) and (7) were found whose SHG efficiencies (1.9μ) were remarkably high at 100 times that of urea and 490 times urea, repetively. These values compare with typical values for p-nitroanilines of 80-160 times that of urea. (Patent applications covering compounds (6) and (7), their manufacture, and their use in NLO applications have been filed by ICI.) The synthetic routes used to access (6) and (7) are shown in Fig.2. The chemistry is standard, the only point of note being the necessity for rigorous purification of DCNP in order to get material

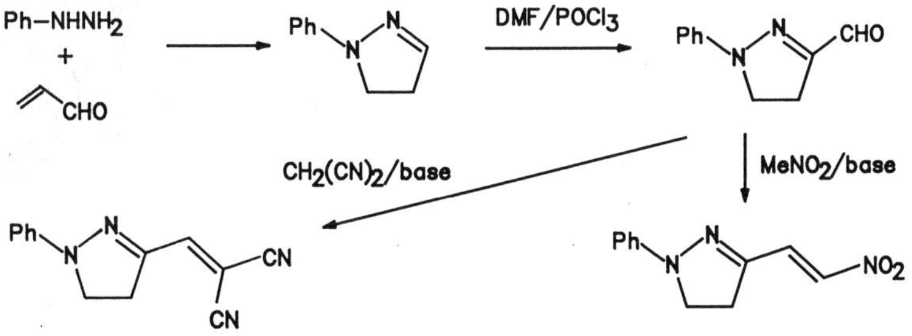

(6) DCNP

SHG 100 x urea

(7)

SHG 490 x urea

Fig.2 Synthesis of 2-pyrazoline derivatives

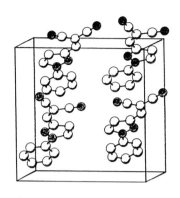

suitable for large crystal growth. An X-ray structure determination of DCNP clearly shows the acentric arrangement of molecules in the unit cell, with adjacent molecules lying parallel to one another (Allen et al 1988a).

Further characterisation of the electro-optic properties of DCNP crystals has been reported elsewhere (Allen 1989).

The discovery of NLO active crystalline forms of DCNP and (7) was a result of pure serendipity. For the synthetic organic chemist, a profound disincentive to preparing new chemicals for crystal growth is the inability to predict <u>ab initio</u> whether a new molecule will crystallise acentrically. Even worse, there is no structure-activity pattern which enables us to use a series of results on homologues to optimise substituent combinations, along the lines that are applied routinely in many other areas of organic chemistry such as pharmaceutical, agrochemical, or colour research. An unpublished programme of research in ICI has involved attempts to calculate how relatively simple organic molecules will pack in a crystal lattice. However, this crystal modelling has only been partially successful, and then only in closely defined areas of chemistry. Another approach that we have tried is a semi-automated survey of molecules which crystallise in acentric space groups, as reported in the Cambridge structure database. Some interesting possibilities were discovered, but clearly this approach is bounded by known structures, and is not amenable to extrapolation.

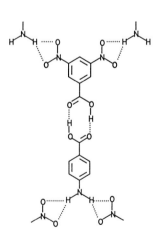

Various so-called crystal engineering approaches have been reported. The inclusion of a homochiral group in a molecule enforces packing in an acentric space group. However, this cannot guarantee NLO activity, since molecular dipoles can still align anti-parallel in an acentric environment, and therefore self-cancel. Hydrogen-bonding effects have also been exploited, and a recent interesting example is provided by Etter et al (1989) where co-crystallisation of p-amino-benzoic acid and 3,5-dinitrobenzoic acid leads to an NLO active crystal (adjacent). However, success can never be guaranteed in such cases.

Yet another approach to inducing acentricity in the crystalline phase is the use of inclusion complexes. Adducts of beta-cyclodextrin were the first reported (Tomaru et al 1984). However this system is limited by the NLO active molecule, and in applying it to various chromogens we have met with little success. Channel complexes of organometallic compounds in thiourea have given a much higher rate of success in achieving acentric arrangements (Tam et al 1989), but again the method appears limited by the size and shape of the guest molecule. A recent report of the activity of p-nitroaniline when occluded in the acentric environment of the zeolite ALPO-5 has revealed a level of SHG activity 37 times that of urea, which is about 10 times better

than previous inclusion complexes (Cox et al 1988). But again the approach
does not appear to be general, as the closely related 2-methyl-4-nitro-
aniline is inactive in the zeolite, although active in the pure crystal.

Further disincentives to the application of crystals in NLO devices include
the specialised techniques, and time, required to grow crystals of
sufficient size, and the difficulty in fabrication of the crystals into
useful devices, in some cases due to their mechanical weakness. For this
reason, the attention of both ourselves and others has been turned to other
orientation techniques.

4. ORDERED POLYMER PHASES

Isotropic polymers appear at first sight to be unlikely hosts for NLO
molecules. However, if the active molecule can be oriented in the
isotropic phase, and then further randomising motion prevented, the
resultant polymer-guest material should retain NLO activity. Such a
material can be obtained by dissolving the NLO species in a polymer;
warming the mixture above the glass transition temperature where the guest
has freedom of movement; applying an external electric field ("poling") to
orient the dipoles of the NLO species; and cooling the mixture back to a
glass, with the field still on until fully cooled, when relaxation to a
fully disordered state is no longer possible.

In practice, this technique leads to polymeric materials which do in fact
have acentric arrangements of NLO species. Unfortunately, these suffer
from the fact that the activity is gradually lost, firstly because motion
is not totally frozen, and residual freedom of the entrapped chromogens
themselves leads to randomisation of orientation. Secondly, relaxation of
the polymer chains allows the chromogens to reorient. As an example,
4-dimethylamino-4'-nitrostilbene (DANS) oriented either in polymethylmeth-
acrylate (PMMA) or polystyrene loses 20% of its initial bulk susceptibility
in 12 hours, and essentially all of it after 10 days at room temperature
(Hampsch et al 1988). A similar situation pertains for 12.5% p-nitroaniline in poly(oxyethylene), whose initial SHG activity of 20-30 times
urea decreases to about the level of urea after 200 hours (Watanabe et al
1988). On the positive side, these doped polymers have good mechanical
strength and a high laser damage threshold, even if the impermanence of the
NLO activity precludes their application in devices.

An improvement in the retention of the required activity can be achieved by
directly binding the chromogen to the polymer backbone. The functionalised
polystyrene (8) still retains ca.30% of its initial SHG activity 3 days
after the warming, poling, cooling sequence (Ye et al 1988). Furthermore,
use of corona poling to induce dipole alignment leads to higher initial and
resultant activities. For example, corona poled methacrylate copolymer (9)
has a SHG effect 3 times higher than direct field poling (Singer et al
1988). Moreover, the activity decreases by only 10% over 30 days, and by
this time has reached a constant level. The corresponding corona-poled
guest-host polymer loses about 75% of its activity over the same period.
Clearly, then, a decrease in the mobility of the active chromogen in the
polymer due to covalent binding can result in stable, NLO active polymer
films.

The synthesis of polystyrene (8) is amenable to variation of the chromogen;
it need only possess an electrophilic site to react with the hydroxylated
polystyrene precursor. Similarly, copolymer (9) is derived from a standard

(8)   (9)

donor-acceptor substituted azobenzene, functionalised at one end with a
methacrylate ester.

If the side chain attaching the chromogen to the polymer backbone is long
enough, and the chromogen itself is mesogenic, the polymer can assume
liquid crystalline properties. In this case, acentric order induced by
poling can be retained by the structure imposed by the liquid crystal
packing, rather than by the rigidity of the glassy polymer matrix (Buckley
et al 1988).

There are many advantages to poled polymer films. The syntheses themselves
are normally straightforward, and the chemist can modify the properties of
the film by varying the nature of the backbone, the chromogen, the type and
size of linkage between backbone and chromogen, and the polymerisation
method. Successful orientation of chromogens is dependent only on a
sufficiently high ground state dipole moment, and enough flexibility in the
system to permit freedom for reorientation. The polymers frequently have
favourable chemical, optical, and mechanical properties, and several
techniques are known for converting them into materials suitable for device
application. All these reasons explain why these materials are assuming
increasing importance in the development of organics-based NLO devices.

5. CALCULATIONS ON CONJUGATING PATHWAYS

Although it is known that there are many chemical possibilities for the
conjugating link between pi-donor and acceptor groups in an NLO species, it
was of interest to determine which was the best. The easiest access to
such a result was by calculation. A series of CNDOVSB calculations was
first run on the molecules $Me_2N-(C_6H_4)_n -NO_2$, where the C_6H_4 groups are
1,4-phenylene (Morley et al 1987c). The hyperpolarisability values were
found to increase to a maximum for n = 8 (β = 54 x 10^{-30} cm^5 esu^{-1}).
However, it was felt that a more realistic measure of the effectiveness of
a molecule is its hyperpolarisability per unit volume (so-called
hyperpolarisability density). According to this parameter, the best path
length was only n = 2 or 3 (bi- and terphenyl derivatives) (Fig 3).
Extending the length increased the volume disproportionately with respect
to β.

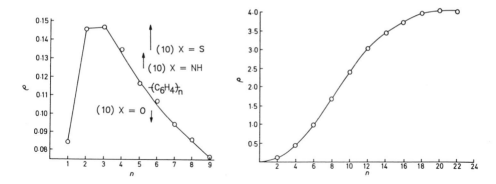

Fig.3   Calculated hyperpolarisability density (ρ) for donor-acceptor
substituted polyaryl (left) and polyene (right) systems.

More recent unpublished studies show that the
same saturation phenomenon occurs for the
correspondingly 2,5-substituted furans,
pyrroles, and thiophenes (10). The thiophene
series (X = S) is calculated to be the best,
and about 3 times more active than the
benzenoids. In this case, the optimum length
is again about 3 thiophene rings.

The situation changes appreciably for substituted polyenes $Me_2N(CH=CH)_n NO_2$.
Their calculated hyperpolarisability density only stops increasing after
about 20 double bonds. The value reached is more than 25 times that of
the polyphenylene derivatives. For maximum NLO effect, it therefore
appears that appropriately substituted polyenes are to be preferred over
polyaryls.

However, two words of caution are in order. Firstly, the calculations
over-estimate both the experimentally measured electronic transition
energies of the polyenes and their ground state dipole moments. By
implication, the hyperpolarisabilities may also be over-estimated,
particularly at long chain lengths. Of more practical concern is the
synthetic access to substituted polyenes, and their ultimate chemical
stability. As the chain length increases, stability falls. It is only
recently that unsubstituted polyenes as long as 15 double bonds have been
isolated and identified (Knoll et al, 1988). However, substitution of
alkyl groups has a stabilising effect, so that synthetic carotenoid
derivatives containing up to 19 double bonds are known (Karrer et al,
1951). Although attracted by the potentially high NLO activity of the
polyenes, we were also concerned by their chemical stability, and for that
reason concentrated our efforts on the hybrid species discussed below.

6. LANGMUIR-BLODGETT FILMS

As an alternative method of forming thin, regular films of acentrically
oriented NLO molecules, we have investigated the Langmuir-Blodgett (LB)
technique. This has been amply described in the NLO literature already, so
we make only a few general comments. Head-to-tail (Z-type) deposition of

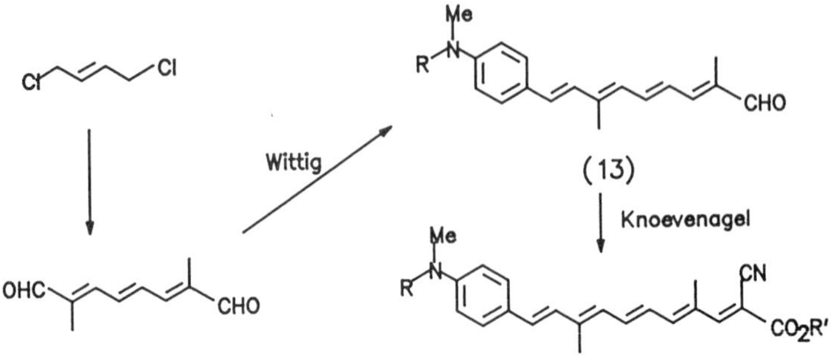

(11) R = n-C$_{12}$H$_{25}$, R' = H

(12) R = C$_2$H$_4$CO$_2$H, R' = n-C$_{12}$H$_{25}$

Fig.4   Synthesis of donor–acceptor substituted polyenes

layers of some molecules are thermodynamically unstable when a normal
amphiphilic arrangement of polar and apolar groups are used. Instead, the
centrosymmetric head–to–head, tail–to–tail (Y–type) deposition is more
favoured. Of course, this will be NLO inactive. In order to overcome this
drawback two different molecules may be used for the odd and even numbered
layers, respectively, in an alternating Y–type arrangement. These two
molecules are distinguishable by being of reversed amphiphilicity, but
based on the same chromogen. The Y–type structure built up can be more
stable, is acentric, and therefore is potentially NLO active.

Earlier work in ICI had concentrated on diarylazo compounds (Ledoux et al,
1987, 1988). In order to achieve higher NLO activity, with an acceptable
level of chemical stability, the amphiphile–reversed donor–acceptor polyene
species (11) and (12) were synthesised, besides several related species
(Fig.4). In (11), the cyanoacetic acid–derived end group acts as the
hydrophilic head, with a dodecyl chain attached to the amino pi–donor being
lipophilic. The situation is reversed in (12) where a carboxyethylamino
donor group is the hydrophilic head. The cyanoacetate group is now
esterified with the lipophilic dodecyl chain. LB monolayers of (11) and
(12) were found to be relatively unstable in air, decolourisation
suggesting either oxidation or electrocyclisation. However, they could be
protected by an inert overcoat.

Characterisation of the LB mono– and multilayer films based on (11) and
(12) has been published elsewhere (Allen et al 1988b), so the results are
only summarised here. The nonlinear susceptibility X$^{(2)}$ of a monolayer of
these compounds is about 3–5 times higher than that of diarylazo compounds.
The hyperpolarisability value, β, of (11) was derived to be 830±70 x 10^{-30}
esu, which is exceptionally high. (The highest published β value we are
aware of is that of the extended quinodimethane (14) for which a β > 1000 x
10^{-30} esu has been reported (Garito et al 1988).) A measured solution
EFISH β value for the aldehyde (13; R = Me) was found to be 159 x 10^{-30}
esu, gratifyingly close to the calculated value of 165 x 10^{-30} esu (Ledoux
and Gordon, unpublished).

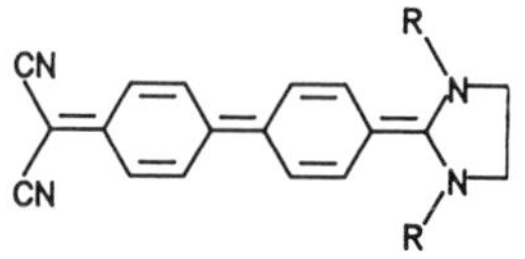

CN

CN

R

R

(14)

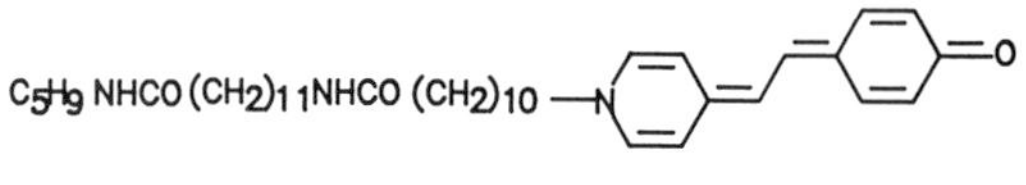

C₅H₉NHCO(CH₂)₁₁NHCO(CH₂)₁₀

(Note: rendering formula) $C_5H_9\,NHCO(CH_2)_{11}NHCO(CH_2)_{10}$ —N〈...〉=〈...〉=O

(15)

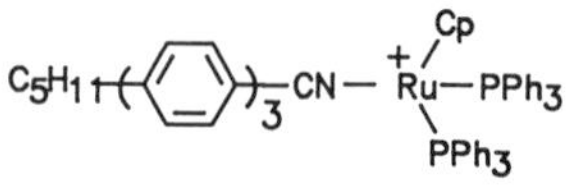

C_5H_{11}—〈( )〉₃—CN—Ru—PPh₃ with Cp and PPh₃

(16)

O_2N—〈 〉—N=N—〈 〉—N(Me)($C_2H_4CO_2H$)

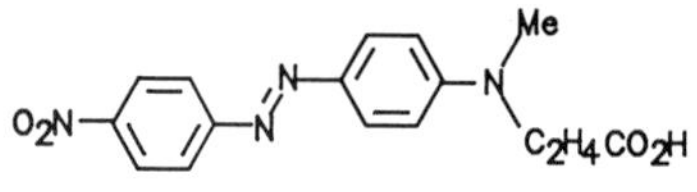

(17)

While LB films have several attractive features as NLO materials, they still have disadvantages:
- the instability in many cases of Z-type structures;
- the necessity to synthesise and deposit two different molecules for alternating Y-type structures;
- structural irregularities both in-plane, and as layers build up;
- the chemical reactivity of exposed monolayers;
- the mechanical stability of films once deposited;
- the time required to build up many-layered structures;
- the question of further fabrication.

The design and synthesis of molecules which overcome some of these problems appears to be a central theme of organic NLO research at present. For instance, stable Z-type multilayer structures have been reported for several molecules which include amide groups in the lipophilic tail (15) (Popovitz-Biro et al, 1988), and for ruthenium complexes of liquid crystals (16) (Richardson et al, 1988). The unconventional bipolar molecule (17), which contains no lipophilic tail, builds up into a Z-type structure, presumably through intermolecular H-bonds between the carboxyl and nitro groups (Barton et al, 1988). Further progress in this area can be expected.

7. CONCLUSION

This review of current progress in the design and synthesis of organic molecules for second order NLO applications has of necessity been brief. However, it should serve to indicate how organic chemists, especially those in ICI, have been thinking about this topic in the recent past. Our current efforts in ICI are in the area of polymeric thin films, where organics-based devices are now being reported, and in the less well developed, but chemically exciting, area of molecules designed to overcome the weaknesses listed above for LB film application.

This review has said nothing of third order NLO materials, which have largely been dominated by conjugated polymers, such as polydiacetylenes. The characterisation of small molecules and attendant substituent effects is not nearly as well developed as in second order NLO studies, although recent patents claim high third order nonlinearities in some cases. This area, and novel conjugated polymers, are also of interest in ICI.

8. ACKNOWLEDGMENTS

Most of the synthetic work in ICI has been carried out by B Bothwell. Characterisation was by colleagues in other parts of ICI, as well as by various collaborators in the UK and France. These collaborations have been supported by JOERS (UK) and ESPRIT (EEC) funding.

9. REFERENCES

Allen S, McLean T D, Gordon P F, Bothwell B D, Hursthouse M B and Karaulov S A 1988a J Appl Phys 64 2583

Allen S, McLean T D, Gordon P F, Bothwell B D, Robin P and Ledoux I 1988b SPIE Proceedings 971 206

Allen S 1989 Organic Materials for Non-linear Optics ed R A Hann and D Bloor (London: RSC) pp 137-150

Barton J W, Buhaenko M, Moyle B and Ratcliffe N M 1988 Chem Commun 488

Buckley A, Calundann G W and East A J 1988 SPIE vol 878 Multifunctional Molecules 94

Cox S D, Gier T E, Stucky G D and Bierlein J 1988 J Amer Chem Soc 110 2986

Docherty V J, Pugh D and Morley J O 1985 J Chem Soc Faraday Trans 2 81 1179

Garito A F, Heflin J K, Wong K Y and Zamani-Khamiri O 1988 Mat Res Soc Symp Proc 109 91

Etter M C and Frankenbach G M 1989 Chem of Materials 1 10

Gusten H, Heinrich G and Fruhbeis G 1977 Ber Bunsenges Phys Chem 51 810

Hampsch H L, Yang J, Wong G K and Torkelson J M 1988 Macromolecules 21 528

Karrer P and Eugstler C H 1951 Helv Chim Acta 34 1805

Knoll K, Krouse S A and Shrock R R 1988 J Amer Chem Soc 110 4424

Kurtz S K and Perry T T 1968 J Appl Phys 39 3798

Ledoux I, Josse D, Vidakovic P, Zyss J, Hann R A, Gordon P F, Bothwell B D, Gupta S K, Allen S, Robin P, Chastaing E and Dubois J C 1987 Europhys Lett 3 803

Ledoux I, Josse D, Fremaux P, Piel J-P, Post G, Zyss J, McLean T D, Hann R A, Gordon P F and Allen S 1988 Thin Solid Films 160 217

Morley J O, Docherty V J and Pugh D 1987a J Chem Soc, Perkin Trans 2 1357

Morley J O, Docherty V J and Pugh D 1987b J Chem Soc, Perkin Trans 2 1361

Morley J O, Docherty V J and Pugh D 1987c J Chem Soc, Perkin Trans 2 1351

Morley J O, Docherty V J and Pugh D 1989 J Molec Electronics in press

Popovitz-Biro R, Hill K, Landau, E M, Lahav M, Leiserowitz L, Sagiv J, Hsiung H, Meredith G R and Vanherzeele H 1988 J Amer Chem Soc 110 2672

Richardson T, Roberts G G, Polywka M E C and Davies S G 1988 Thin Solid Films 160 231

Singer K D, Kuzyk M G, Holland W R, Sohn J E, Lalama S J, Comizzoli R B, Katz H E and Schilling M L 1988 Appl Phys Lett 53 1800

Tam W, Eaton D F, Calabrese J C, Williams I D, Wang Y and Anderson A G 1989 Chem of Materials 1 128

Tomaru S, Zembutsu S, Kawachi M and Kobayashi M 1984 Chem Commun 120

Watanabe T, Yoshinaga K, Fichou D and Miyata S 1988 Chem Commun 250

Ye C, Minami N, Marks T J, Yang J and Wong G K 1988 Mat Res Soc Symp Proc 109 103

Inst. Phys. Conf. Ser. No 103: Section 2.1
Paper presented at Int. Conf. Materials for Non-linear and Electro-optics, Cambridge, 1989

Growth of organic single crystals

Norbert KARL

University of Stuttgart, 3. Phys. Inst., D-7000 Stuttgart-80, Fed. Rep. of Germany

ABSRTRACT: An overview is given on classical and more recent methods which can be used for growing single crystals of small molecular weight organic molecules. Influences of impurities and dopant molecules on crystal growth, habit and quality, and problems arising with growth of homogeneous mixed crystals are briefly discussed together with techniques useful for purity and quality assessment, crystal orienting, processing and handling.

1. INTRODUCTION

Growth of high quality organic single crystals has found a renaissance following the discovery of some organic compounds with very high optical nonlinearities which far exceed those of inorganic materials, see e.g. Zyss (1985), Chemla and Zyss (1987), Günter and Huignard (1988), Hann and Bloor (1989). An overview is given on the "classical" crystal growth methods which have been used to obtain big single crystals of small molecular weight organic compounds (mainly aromatics) for the investigation of their electrical and optical properties and their triplet spin states. Solution growth (by either continuously decreasing temperature or solvent volume), melt growth (by directional cooling after Bridgman-Stockbarger) and sublimation growth are covered.

More recent promising developments in the organic field are melt growth by seeding and pulling techniques (Nacken-Kyropoulos and Czochralski), near to isothermal growth from an undercooled melt (where the growing crystal is always the warmest part), flux growth from binary or more component systems, and growth into optical wave guiding capillaries. There is little ecperience in these techniques and they merit further studies.

Growth of epitaxial organic crystal layers, either under near thermodynamic equilibrium conditions, or by molecular beam epitaxy (MBE) is shown to be a promising field. This may eventually lead to a hybrid technology with integration of *organic* materials into electronic, optoelectronic, and nonlinear optic micro-circuits on inorganic substrates.

Growth under different purity conditions or doping levels frequently leads not only to different crystal habit, but also to widely varying crystal quality. Better knowledge of the microscopic interactions and growth mechanisms is therefore highly desirable.

Dopant acceptability and conditions for the formation of homogeneous mixed crystals are of fundamental importance for tailoring special crystal properties. In this connection, growth procedures which can yield doped or mixed crystals with spacially constant or deliberately adjustable concentration profiles require special attention.

Finally, a brief review of characterization techniques is given, which allow a quantitative classification of the content of physical and chemical defects together with some possibilities for crystal orientation, cutting, polishing, and handling.

2. CLASSICAL METHODS FOR THE GROWTH OF *ORGANIC* SINGLE CRYSTALS

The main crystal growth methods which have been used "classically" for producing single crystals of (non-polymeric) organic materials are: solution growth, melt growth, and sublimation growth. For a general review on these methods the reader may be referred to the work by Smakula (1962), Laudise (1970), Wilke (1973 and 1988), Chernov (1984), and Sloan and McGhie (1988).

Solution growth of optically nonlinear organic crystals has recently been described by Badan et al. (1987). In solution growth it is usually indispensable to start from a saturated solution which is free of spontaneously formed "parasitic" seed crystallites and foreign nucleation centers, such as dust or scratches at the container wall, and to immerse a high quality seed crystal. It is also important in most cases to generate a deliberate small undersaturation, no more than just enough to dissolve a minor fraction of the seed crystal and thus clean the surfaces from irregular seeding centers immediately before the growth period.

With the *evaporation method* slow evaporation of the solvent can be accomplished by placing a beaker with the solution into a vacuum-desiccator and letting a dry inert gas such as nitrogen stream through the desiccator under well controlled flow and constant ambient temperature conditions. A surrounding large volume water jacket is useful to stabilize against short time temperature fluctuations.

Alternatively, the *cooling method* may be applied. In this method a closed container is best immersed into a thermostated bath with a viewing window whose temperature is first slightly raised to remove parasitic seeds and subsequently decreased at a controlled rate. The *volume* growth rate will depend on the slope of the solubility versus temperature curve, which is usually not independent of temperature. Moreover, since the surface area increases during growth and the individual crystal faces grow at different speeds, maintaining constant growth speeds of the face normals may require specifically designed temperature programming.

More sophisticated versions of solution growth by cooling are based on a continuous dissolution of the starting material at a hotter part of the growth apparatus and crystallization at a slightly colder *constant temperature* place. Several technical suggestions may be found in the book by Wilke (1988).

It is frequently crucial to use an appropriately cut, high pefection seed crystal because crystallographically different faces usually grow with different velocity and quality, and display differences in their defect structures and impurity concentrations.

The presence or absence of certain impurities may "catalyze" certain growth mechanisms and hence greatly influence the final crystal quality (see section 5).

Stirring of the growth solution, and/or intermittent or reciprocated rotation of the crystal in order to obtain a thorough equalization of concentration and temperature is another point which needs careful consideration. If performed sufficiently gently, it can greately improve the growth results. Further details and general guidelines on solution growth are given in the article by Hooper et al. (1980).

Growth of organic crystals from the melt by *directional cooling* after Bridgman and Stockbarger has been described by a number of authors; these investigations are reviewed by Sloan and McGhie (1988). This method, for which a useful growth speed usually lies between 0.1 and 1 millimeters per hour, suggests itself at least for a first trial, whenever a material can be kept molten without thermal decomposition over several days. The material may require a certain purity level to achieve this. A conical pyrex glass ampoule is frequently chosen, which can be evacuated and sealed by a flame. The shape of the seeding region of the ampoule, the temperature gradient and the shape of the isothermals at the melt/crystal interface play an important role in determining the final crystal quality. Some suggestions may be found in the paper by Sherwood and Thomson (1960) and in the article by Karl (1980). A variant ("reflector furnace") which has proved very useful in practice is that invented by Gerdon (1973). The ampoule is heated in an evenly wound tubular glass oven. The placing of a gilded, semitransparent glass cylinder around the oven reflects the heat back to the ampoule over a certain fraction of its length and creates a melt. Movement of the cylinder relative to the oven moves the liquid/solid interface. An example of a Bridgman-grown single crystal is given in Fig.1

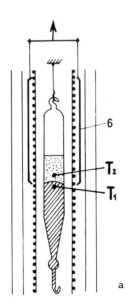

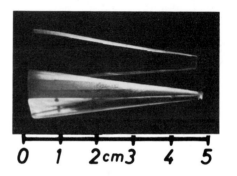

Fig. 1 Bridgman growth variant with moving reflector oven, schematic drawing (a) and photograph of a cleaved 2,3-dimethylnaphthalene crystal grown by this method (b).

Sublimation growth of *large, highly perfect* organic single crystals from a limited amount of starting material is difficult and has not often been investigated or demonstrated. Smaller crystals, however, up to $\sim 1\,cm^3$ can be obtained readily by the plate sublimation technique (Karl 1971, Karl 1980) 1980). This method uses a cylindrical evacuated glass ampoule which is placed between two massive, evenly heated metal plates kept at slightly different temperatures ($\Delta T \sim 0.5°$). Lateral heat flow through the cylindrical periphery is compensated by a surrounding wire-wound glass cylindrical heater. Seed formation can be induced by exceeding the stable Ostwald-Miers supersaturation regime by adjusting a slightly larger temperature gradient, and the average number of seeds can be controlled by the duration of this additional supersaturation. The material sublimes along the one-dimensional temperature gradient from one of the large planar faces of the ampoule to the other, as shown in Fig. 2 .

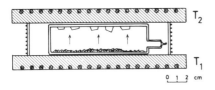

Fig. 2 Plate sublimation method, principle and some examples of single crystals grown by this technique.

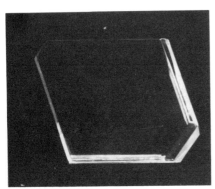

anthracene, $\sim 7 \times 6 \times 1$ mm^3

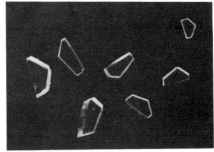

anthracene, $\sim 1.5 \times 0.5 \times 0.3$ mm^3

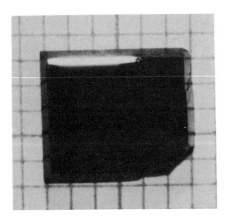

donor:acceptor complex TTF:CA (tetrathiafulvalene:chloranil), $6 \times 5 \times 1.5$ mm^3

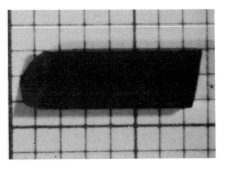

donor:acceptor complex   PTZ:TCNQ (phenothiazine:tetracyanoquinodimethane) $6 \times 2 \times 2$ mm^3

Owing to its inherent high degree of temperature homogeneity, the plate sublimation technique can not only yield highly perfect single crystals, but is also especially well suited to growing single crystals of weakly bound molecular complexes, such as stoichiometric donor:acceptor complexes, where larger temperature differences often lead to a segregation of the components. Growth in a low pressure inert gas (1 - 400 torr N$_2$) can sometimes influence the formation of thin platelets or long thin needles instead of bulky crystals.

3. TRIALS WITH UNCONVENTIONAL METHODS

Isothermal *solution growth* is also possible by reducing the solubility using slow dilution of the solvent with one of lower solubility. This can be accomplished either by placing a second flask with an orifice of suitable diameter containing the volatile dilutant, into the chamber containing the saturated solution. The chamber can conveniently be a desiccator. Alternatively, a poorer solvent can be caused to diffuse into the solution through a semipermeable membrane. Following suggestions by Vaala (1973), large ß-carotene crystals have been obtained by this latter method.

Melt growth methods which are based on seeding and pulling techniques (Nacken-Kyropoulos and Czochralski) cannot easily be applied on small molecular weight organic compounds because these usually develop a considerable vapor pressure at their melting point, leading to problems of heteronucleation and "snowing". The technique is suitable for selected compounds which in relation to their molecular weight have exceptionally low melting points (e.g. as a consequence of low molecular symmetry, little polarity, mobile side groups or unfavorable shape) and thus have only small vapor pressure at their melting point such as benzophenone, benzil or salol (Bleay et al. 1978). Alternatively the provision of a closed chamber growth environment with negligible temperature differences between any parts which are in contact with the vapor has been successfully developed in an apparatus described elsewhere (Karl, 1989). The basic principle of Czochralski growth and examples are given in Fig. 3.

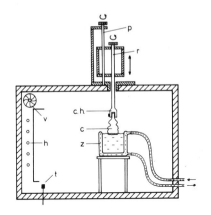

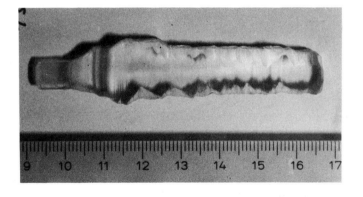

Fig. 3
Czochralski
growth, prin-
ciple, salol
crystal during
growth and
benzophenone
crystal.

Growth of organic crystals from undercooled melt has great advantages as demonstrated by Klapper and coworkers (Scheffen-Lauenroth et al. 1981, Klapper 1984). Organic melts frequently allow considerable undercooling without spontaneous nucleation, if clean conditions are maintained. Immersion of a seed leads to crystallization under release of the latent heat of solidification, which in turn raises the temperature at the growing crystal / melt interface up to the melting point. From then on, crystallization speed is governed by the rate of heat removal *through the melt* by conduction and convection. This can simply be adjusted by keeping the surrounding temperature slightly (tenths of degrees) below the melting point and, if additionally necessary, thermally insulating by providing a sufficiently large melt volume. Since the crystal is immersed into its melt, gravitational forces are balanced out to a great extent allowing very large crystals of up to several 100 g to be produced, e.g. of benzophenone, benzil and salol (Klapper 1984); examples are shown in Fig 4. Crystals grown by this near to isothermal method are essentially strain-free and tend to display very low dislocation contents, as demonstrated by Scheffen-Lauenroth et al. (1981).

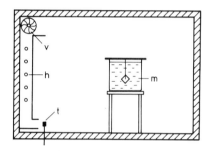

Fig. 4 Growth from an undercooled melt, principle, growing benzophenone crystal, and examples of crystals grown by this method.

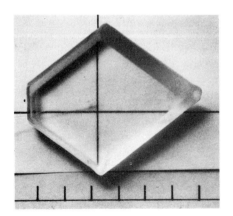

salol   benzophenone

Flux growth is a crystal growth method which uses an auxiliary component which is solid at room temperature, but acts as a solvent at elevated temperature. This component can be a foreign molecule which has only the function to lower the melting point, or it can be a component of the multicomponent growth crystal added in great excess above the stoichiometrically necessary concentration. An organic example is the growth of the 1 :1 donor:acceptor complex between naphthalene and pyromellitic dianhydride (PMDA) by lowering the ampoule in a Bridgman type setup with intermittent rotation of the ampoule (Karl 1988). Knowledge of the phase diagram is usually required, which can be obtained by differential scanning calorimetry, DSC. This hinders the use of flux growth as a quick routine trial method for new systems; but it should not be an obstacle if production quantities of organic crystals are required. It must be stressed, however, that the auxiliary component, if it is not a stoichiometric and therefore wanted constituent, may be found incorporated in in the desired crystal in a small concentration.

Crystal growth into optical wave guiding *capillaries* is a promising technique for making practical use of optically highly nonlinear organic crystals for efficient second harmonic generation and frequency mixing in cheap devices, cf. Badan (1987). Although remarkable progress has been made recently by Holdcroft et al. (1988), achievement of physical homogeneity of the core and constancy of the crystallographic orientation over a certain required length are still problems. A method which has proved very useful for obtaining void-free zone refining ingots, the so-called "reverse zone refining technique" (cf. Karl 1980) can be of use here. Briefly, it consists of moving a periodic array of small heater coils across the length of the zone refining tube (or capillary) which has one closed end (e), as shown in Fig. 5, and has been more or less tightly filled under vacuum from this end on. The other end is sealed after filling at a point sufficiently far away from the organic material. Thus an expansion volume is left. Movement is accomplished in the *reverse* direction (in comparison with *normal* zone refining), i.e. the molten zones enter the capillary and ingot at the closed (filled) end. Melting of organic molecular crystals causes a volume increase (by typically 10 or 20 %) which leads to a compressive stress. Voids are compressed and, as soon as this is reached, all solid sections of the ingot will have to move in order to yield to the pressure. This happens in the case of short sections, whereas if the sections are too long (e.g. if only one small molten zone is produced and pulled across the entire ingot several times) fracture occurs because liquids are fairly incompressible.

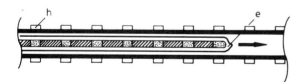

Fig. 5   Principle of the reverse zone refining technique, suggested for the growth of waveguiding single crystal fibers into capillaries.

4. EPITAXIAL ORGANIC THIN FILMS

Epitaxial organic thin films on inorganic substrates, such as e.g. silicon, may constitute another possible access towards integrated optic, nonlinear optic and electrooptic devices. At present research is concentrated on investigating the basic interaction and epitactic crystal growth mechanisms (Kobayashi et al. 1983, Yase et al. 1984, Karl 1988, Möbus et al. 1989). Problems with the differences of the lattice symmetries and the sizes of the lattice meshes between most organic molecular crystals and conventional inorganic sustrates like silicon, as well as problems with the different thermal expansion coefficients will have to be solved.

Moreover, it appears necessary to gain a detailed physical and chemical under-standing of the microscopic interactions that lead to heteroepitaxy. As an exam-ple, preliminary results obtained with vacuum vapor deposition of perylene-tetracarboxylic-dianhydride (PTCDA, a commercial very stable organic dye pig-ment) on freshly cleaved and vacuum-heated NaCl faces is worth mentioning. In one of several possible orientations, the PTCDA molecules stand on their carbo-nyl oxygen atoms of their anhydride ends, upright on the NaCl (100) plane with a remarkably small lattice misfit of the oxygen−Na^+ interactions (Möbus et al. 1989, see also Karl 1988), see Fig. 6. On UHV-prepared Si(111) 7×7 reconstructed surfaces, a chemical bond to the carbonyl oxygen is indicated by near edge X-ray absorption fine structure (NEXAFS) spectra (Zimmermann et al. 1989).

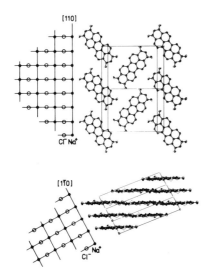

Fig.6   Epitaxial growth of PTCDA (perylene-tetracarboxylic-dianhydride) on the (001) cleavage plane of NaCl; the relative orientation was obtained from HV transmission electron diffrac-tion patterns of a removed film.

5. INFLUENCES OF IMPURITY AND DOPING LEVELS ON GROWTH

Crystal growth, especially of organic molecular crystals can be strongly cata-lyzed or blocked by chemical impurities which may be present from the starting material, or deliberately added as dopants, see e.g. Addadi et al. (1985). These molecules may cover the growing surface and completely alter the kinetic balance of adsorption and desorption during growth, in that they block the favorable addition places or surface seed formation centers or, alternatively act as prefer-red nucleation centers. The first situation is likely if the geometric shape of a foreign molecule and of a possible adsorption place for a regular molecule are complementary in that they fit like key and lock. The latter situation is expected if for example a moiety of a larger molecule can fairly well replace a regular molecule. The other end will stick out of the surface and may then act as a nucleation center for a new molecular layer. This seems to hold for tetracene impurities in anthracene. The entire field is barely investigated and offers many interesting aspects for further research. It should be possible not only to improve crystal quality, but also to tailor and optimize growth parameters such as growth speed and crystal habit.

6. GROWTH OF DOPED AND MIXED CRYSTALS

Poor dopant acceptability and growth with non-constant doping concentration are the two major difficulties one is confronted with if one tries to grow doped or mixed crystals.

Dopant acceptability of organic molecular crystals is frequently very small and the formation of homogeneous mixed crystals over a considerable range of concentrations is the exception rather than the rule. As Kitaigorodski (1984) has pointed out in some detail for purely van der Waals-bonded solids, a fairly close geometrical fit of the molecule to be incorporated into a given host crystal lattice with the (unrelaxed) vacancy formed on removing a host molecule is necessary to balance the energy needed for removal by the the energy gained on binding the new molecule. These binding energies, as has been proved in many examples, are essentially determined by the number of close atom–atom contacts and can be calculated on the basis of empirical atom–atom potentials. Geometrically unfavorable or larger molecules would have to squeeze away other neighbors which costs energy and therefore reduces the total lattice binding energy.

For a more quantitative estimate, imagine an optimal geometrical superposition of the dopant molecule with one molecule in the host lattice and calculate the ratio of the volumes of the residual non overlapping molecular parts and the overlapping parts between these two. If this ratio exceeds $\sim 20\%$, a considerable incorporation on regular lattice sites will essentially be prohibited. It should, however, be taken into account that it may be energetically more favorable to incorporate dopant molecules in combination with other lattice defects, such as dislocation lines, and that they may therefore appear with altered physical properties, e.g. with shifted spectral lines.

To give examples, wide range miscibility was obtained in our laboratory for the systems with very similar molecular sizes of 2,3-dimethylnaphthalene/anthracene, 2,3-dimethylanthracene/tetracene and brazan/tetracene, which were analyzed by differential scanning calorimetry (DSC), (Karl 1984, Karl et al. 1985).

As to the dopant or, in mixed crystals, second component concentration (c) profiles, one should be aware of the fact that a crystal (s) growing into a melt or solution (1) phase will usually form at a different composition, $c_s = k\,c_l$, ($k \neq 1$), which is dictated by the (general) non-coincidence of the solidus and liquidus lines in the phase diagram. Therefore, in a limited volume the composition ratio in the nutrient phase, and as a consequence, in the crystal will change during growth. In solution growth with the temperature lowering method, a further complication arises if the distribution coefficient $k = c_s/c_l$ is temperature dependent.

Doped or mixed crystals with constant concentration can be obtained either if the nutrient phase is present in great excess (i.e. if only a small fraction of the source material is transferred into crystals), or if the composition of the nutrient phase is kept constant artificially. The latter is automatically the case in near to equilibrium (i.e. non kinetically limited) sublimation growth from separate sources kept at the same temperature. However the concentration ratio can then be varied only within narrow limits, by selecting the absolute temperature within the range useful for growth and then only if the two components have different heat of evaporation and hence different slope of the vapor pressure versus temperature curves. Non thermodynamic equilibrium between the component sources and the mixed crystal as in the case of molecular beam epitaxy constitutes a convenient possibility for obtaining doped or mixed crystals with a spacially constant concentration at a level freely selectable within the limits of homogeneous miscibility.

7. CHARACTERIZATION OF CRYSTAL QUALITY AND PURITY

Crystal quality and purity may be crucial parameters for successful applications, especially in nonlinear optics, and therefore require quantitative specification. X-ray diffraction—vector distribution functions (rocking curves) can give a measure of single crystallinity with spacial resolution of about 1 mm . Sufficiently perfect crystals can be further analyzed by X-ray topography, see e.g. Scheffen-Lauenroth et al. (1981) In rare cases, macroscopic volumes (of the order of 1 cm^3) can be found free of dislocations, even in weakly van der Waals bound organic molecular crystals. 2,3-dimethylnaphthalene, as shown in Fig. 7 is an example (cf. Karl 1984). Optical methods of quality assessment are : reflection interference contrast microscopy of surfaces, Fig. 8, monochromatic conoscopic transmission interference fringe patterns ("optical axes figures") of plane parallel plates in convergent light between crossed polarizers, Fig. 9, and monochromatic Fabry-Perot fringe patterns in reflected light, Fig. 10. High crystallographic quality is generally necessary in nonlinear optics applications for obtaining long phase matching coherence lengths.

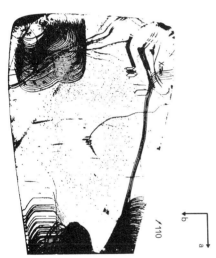

Fig. 7 X-ray topogram of a 2,3-dimethylnaphthalene crystal slice cleaved from a Bridgman grown crystal similar to the one shown in Fig. 1b. The dislocation lines in the top left corner were introduced during cleavage; those in the lower part originate from defects in the seeding region. Apart from several dislocation loops, which may stem from handling, the major part of the sample is free of dislocations.

Fig. 8 Reflection interference contrast micrograph of a sublimation grown biphenylene : pyromellitic-dianhydride complex crystal showing growth hillocks which are probably associated with a screw dislocation growth mechanism.

Chemical purity is not only a problem for obtaining reproducible crystal growth conditions, but also plays an important role for reducing unnecessary optical absorption losses which in addition to increasing the optical loss may lead to local heating and eventual destruction of the crystal.

Purification procedures and purity assessment, although frequently important prerequisites, will not be discussed here, because these have been reviewed in detail elsewhere. The reader is refered to the literature (Karl 1980, 1981, 1984, Sloan

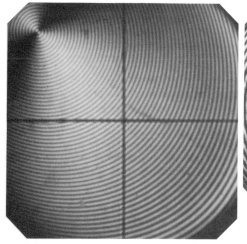

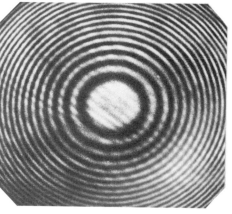

Fig. 10 Fabry-Perot interference fringe pattern of a
cleaved 2,3-dimethylnaphthalene
crystal slice, obtained in reflected monochromatic light
(λ = 589 nm), polarized in one
of the principal planes of the indicatrix. These fringes are el-
liptic, because the index of refraction changes anisotropically with the angle of incidence.

Fig. 9 Monochromatic, conoscopic interference fringe pattern of a (001) anthracene crystal plate in convergent light between crossed polarizers, diagonal
position; the clear appearance
of the interference fringes is
an indication of a high crystallographic perfection.

and McGhie 1988). High pressure liquid chromatography, HPLC, however, is specially mentioned here as one very efficient method for separating impurities. It
can compete with zone refining in cases where the melt/solid distribution coefficients are unfavorable, as in the example given in Fig. 11, and is needed whenever
thermal stability at the melting point is unsatisfactory. This has been the case
for most of the presently investigated organic optically highly nonlinear materials.

8. CRYSTAL ORIENTATION, CUTTING, POLISHING AND HANDLING

Crystal orientation is best done by Laue forward scattering. Live observation of
Laue patterns during rotation of the crystal to be oriented is an indispensable aid
for a quick and economic orientation procedure with low symmetry organic crystals. Comparison of the observed Laue pattern for a certain crystal orientation
with the computer-generated theoretical pattern is usually necessary for a final
proof of the correct interpretation of the Laue photograph of an arbitrary direc-
tion(Karl et al. 1981, Karl 1984).

Cutting, polishing, handling, and even storing of soft organic crystals may require
special precautions. Pure, high perfection organic molecular crystals often display a strong propensity towards gliding. These topics have been treated elsewhere (Karl 1980). Satisfactory optical quality polished surfaces of soft organic

crystals has not generally been obtained. This field requires further research,
with for example round-grain polishing abrasives like garnet. An non-negligible

room temperature vapor pressure may soon deteriorate unprotected surfaces; photo-oxidation acts in the same direction. A need for searching alternative technical solutions which do not necessarily require highest quality primary optical surfaces of the organic material seems to be indicated.

ACKNOWLEDGMENT: The author's and his coworkers' contributions to the field described have been supported by the Deutsche Forschungsgemeinschaft (SFB 67 and SFB 329) and by the Stiftung Volkswagenwerk.

REFERENCES

Addadi, L., Berkovitch-Yellin, Z., Weissbuch, I., van Mill, J., Shimon, L.J.W., Lahav, M., and Leiserowitz, L., (1985) *Angew. Chem. Int. Ed. Engl.* **24**, 466

Badan, J., Hierle, R., Perigaud, A., and Vidakovic, P., (1987) in Chemla and Zyss (1987) opus cit. pp. 297-356

Bleay, J., Hooper, R.M., Narang, R.S., and Sherwood, J.N. (1978) *J. Crystal Growth* **43**, 589

Chemla, D.S., and Zyss, J., (eds.) (1987) *Nonlinear Optical Properties of Organic Molecules and Crystals*, Vol. I and II, (Orlando: Academic Press)

Chernov, A.A. (ed.), (1984) *Modern Crystallography III, Crystal Growth* (Berlin: Springer Verlag)

Gerdon, M., Stuttgarter Kristallabor, (1973), published in ref. Karl (1980), p. 51

Günter, P., and Huignard, J.-P., (1988) *Photorefractive Materials and Applications I* (Berlin: Springer)

Hann, R.A., and Bloor, D., (eds.) (1989) *Organic Materials for Nonlinear Optics* (London: The Royal Society of Chemistry)

Holdcoft, G.E., Dunn, P.L., and Rush, J.D. in: Hann and Bloor (1989), opus cit. p.412

Hooper, R.M., McArdle, B.J., Narang, R.S., and Sherwood, J.N., (1980) in: *Crystal Growth* (Pamplin, B.R. ed.) 2nd ed. (Oxford: Pergamon Press), pp. 395-420

Karl, N., (1980) in: *Crystals, Growth, Properties and Applications*, (Freyhardt, H.O. ed.) Vol. **4**, (Heidelberg: Springer Verlag) pp. 1 - 100

Karl, N., Port, H., and Schrof, W., (1981) *Mol. Cryst. Liq. Cryst.* **78**, 55

Karl, N., (1981) *J. Crystal Growth* **51**, 509

Karl, N., (1984) *Materials Science* (Poland) **10**, 365

Karl, N., Heym, H., and Stezowski, J.J. (1985) *Mol. Cryst. Liq. Cryst.* **120**, 247

Karl, N., (1988) *Mol. Cryst. Liq. Cryst.*, accepted for publ.

Karl, N., (1989) *J. Crystal Growth,* to be published

Kitaigorodsky, A.I., (1984) *Mixed Crystals* (Berlin: Springer Verlag)

Kobayashi, T., Fujiyoshi, Y., and Uyeda, N. (1983) *J. Crystal Growth* **65**, 511

Klapper, H., (1984), *Berichte Rhein. Westf. Techn. Hochschule Aachen* **1**, 32

Laudise, R.A., (1970) *The Growth of Single Crystals* (Prentice Hall)

Möbus, M., Schreck, M., and Karl, N., (1989) *Thin Solid Films,* accepted for publ.

Scheffen-Lauenroth, Th., Klapper, H., and Becker, R.A., (1981) *J. Crystal Growth* **55**, 557

Sherwood, J.N., and Thomson, S.J., (1969) *J. Sci. Instr.* **37**, 242

Sloan, G.J. and McGhie, A.R. (1988) *Techniques of Melt Crystallization* (New York: Wiley-Interscience)

Smakula, A., (1962) *Einkristalle - Wachstum, Herstellung und Anwendung* (Berlin: Springer Verlag)

Wilke, K.-Th., (1988) *Kristallzüchtung* (Frankfurt: Harri Deutsch Verlag); revised and updated ed. of the 1973 version (Berlin: VEB Deutsch. Verl. d. Wissensch.

Yase, K., Okumura, O., Kobayashi, T., and Ueda, T. (1984) *Bull. Inst. Chem. Research, Kyoto Univ.* **62**, 242

Zimmermann, U., Schnitzler, G., Karl, N., Umbach, E., and Dudde, R. (1989), *Thin Solid Films,* accepted for publ.

Zyss, J., (1985) *J. Molecular Electron.* **1**, 25

Inst. Phys. Conf. Ser. No 103: Section 2.1
Paper presented at Int. Conf. Materials for Non-linear and Electro-optics, Cambridge, 1989

119

The non-linear optics and phase matching locus of 4-nitro-4'-methylbenzylidene aniline (NMBA)

R T Bailey, G H Bourhill, F R Cruickshank, S M G Guthrie, G W McGillivray, D Pugh, E E A Shepherd, J N Sherwood, G S Simpson and C S Yoon

The Department of Pure and Applied Chemistry, The University of Strathclyde, 295 Cathedral Street, Glasgow G1 1XL

ABSTRACT: Large single crystals ($5 \times 4 \times 1 cm^3$) of the non-centrosymmetric, monoclinic, (m), form of the organic crystal 4-nitro-4'-methylbenzylidene aniline have been prepared by seeded growth from supersaturated solutions. Optically clear specimens, with faces parallel to the (100) and (010) directions, were examined by the Maker fringe technique. These sections showed no dispersion of the dielectric axes with wavelength in the range 440-630 nm. Refractive indices at 1064 and 532 nm are reported and the $\chi^{(2)}$ elements d_{11} and d_{33} were found to be 311 and 3 relative to quartz d_{11} respectively. Type II phase matching was observed at and near normal incidence to the (100) plane at 1064 nm.

1. INTRODUCTION

4-nitro-4'-methylbenzylidene aniline (NMBA) is one of a series of compounds being investigated in the search for organic electro-optic materials of potential use in non-linear optical devices. It is now possible to predict[1], synthesise and evaluate by powder techniques[2] structures of likely importance in this field. Organic, molecular crystals of ultra-high purity and low defect density are necessary to allow the theories of structure - property relationships to be tested. We have developed techniques for the preparation of such crystals[3].

NMBA crystallises in two polymorphic forms[4]. One of these (triclinic P1) is stable at up to 338K and centrosymmetric. Consequently this exhibits no second order non-linear optical properties. The second form (monoclinic, space group Pc, point group m in the standard setting with the mirror plane orthogonal to the b-axis) is non-centrosymmetric and stable to its melting point. On melting and re-freezing to room temperature, this latter form is produced and remains stable for years without reverting to the triclinic form. The monoclinic form can also be recrystallised from solution at room temperature. A detailed report on this phase behaviour will appear later[5]. In the monoclinic form the molecules are planar and lie with the molecular dipoles ~$40°$ apart.

EXPERIMENTAL

The impurity level of the "as received" material, synthesised from
p-nitrobenzaldehyde and p-toluidine, was 2%. The material was
recrystallised three times from purified n-hexane and zone refined with 120
passes (2cm zone) reducing the impurity levels to 150ppm. Large single
crystals were grown from the melt by a modified Bridgman technique, but
were usually strained. This arose from the decomposition of the material in
the melt and subsequent incorporation of the resulting impurities in the
crystal. Although this problem could be minimised by careful temperature
control, growth from toluene solution and ethyl acetate solution yielded
better quality crystals. This last yielded the best quality crystal[6] when
the temperature was lowered by 0.1° C per day. Resulting specimens were
typically 50x30x5 mm^3 in four weeks.

Fig. 1 shows the indexed faces of the crystal and its growth habit (the
standard setting of the crystal is used to define the Miller indices). Fig. 1
unambiguously defines the orientation of the optical flats used in obtaining
the data discussed below.

Fig. 1 The indexing and
habit of NMBA. Q is the
resultant molecular vector
and n_x, n_z are the
dielectric axes.

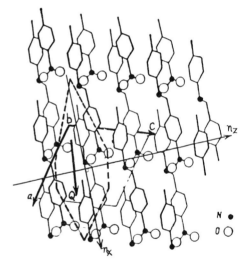

Contrary to previous reports[7], the observed morphology is predicted by
the Hartman theory[8] that the growth rate is proportional to the attachment
energy. Since morphology is determined by the slowest growing faces, the
observed faces should be those of lowest attachment energy.

Previous indexing of the crystal is in error[4]. The correct indexing
requires both that the a- and b-axes must be interchanged and that the
{102} faces be indexed as {101} faces. In this corrected indexing; a =
0.7419, b = 1.1679, c = 0.7447nm and β = 110° 35'. The distinction between
(100) and (001) has been carefully verified by Laue transmission patterns.
Further confirmation of the orientation was obtained from reflectance
spectra acquired on a "Bomem", DA3 Fourier transform spectrometer. The
radiation at 20,000cm^{-1} was polarised in the incident plane and reflected
from the crystal (010) plane. The resultant vector of the two molecular

long axes (Fig. 1) was labelled 0° and the relative intensity of reflected light was recorded as a function of the angle between the incidence plane and this direction. Clearly (Fig. 2), the highest loss occurred when this angle was 0° thus confirming that this was indeed the most polarisable direction of the molecules and thus was closest to the X-dielectric axis as indicated in Fig. 1.

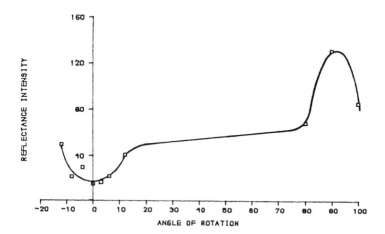

Fig. 2.   The reflected intensity at 20,000cm⁻¹ for in-plane rotation of the (010) face as a function of rotation angle with the polarisation plane parallel to Q at 0°.

OPTICAL CHARACTERISATION

NMBA is a monoclinic crystal of point group *m*.   The d_{ij} matrix is of the form;

$$
\begin{array}{l}
\text{XXX-X-}\\
\text{---X-X}\\
\text{XXX-X-}
\end{array}
$$

Two samples were subjected to Maker fringe analysis in the equipment described previously[2], but arranged for single crystal work.   These samples were indexed (100) and (010) as above (Fig. 1) and were aligned for rotation with their dielectric planes or, where possible axes, parallel to the rotation axes.   All possible Maker fringes were collected from the (010) face, each of three input polarisations being analysed (horizontal, vertical and at 45°) in both possible output polarisations (Fig. 3).

The fringes in the (100) face were obscured by intense phase-matching peaks of closely theoretical (sinc²) shape[6].   Rotation about the Y-axis revealed one phase matched peak at very near normal incidence (Fig. 4). For rotation about the Z-axis, phase matching peaks were observed at -8° and +4°.   At power fluxes up to 1MW cm⁻² the SHG power varied linearly with the square of the incident flux at 1064nm.

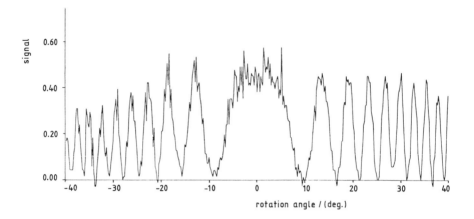

Fig. 3. A typical set of Maker fringes obtained with both 532 and 1064nm light polarised parallel to the Z-dielectric rotation axis. The signal is relative to the maximum intensity generated by d_{11} of quartz.

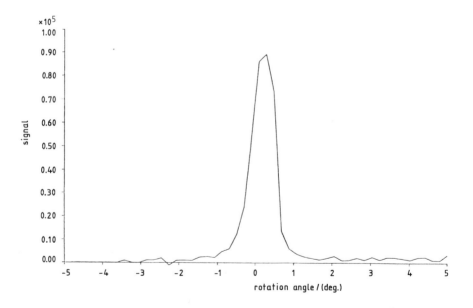

Fig. 4. Type II phase matching in NMBA (100). 532nm light is parallel to the Y-dielectric rotation axis. Incident light at 1064nm is polarised at 45° to the rotation axis. The signal is relative to the maximum intensity generated by d_{11} of quartz.

The linear refractive indices for 1064nm and 532nm light, are required to interpret the Maker fringes, predict the phase matching locus, find the optical axes and to convert the second harmonic intensity, calculated at normal incidence relative to the maximum generated by d_{11} of quartz, into the d_{11} and d_{33} values respectively. These refractive indices were obtained as follows. A prism was cut with its axis along the crystallographic b-axis (i.e. the Y-dielectric axis). One face of the prism was the $(10\bar{2})$ and the apex angle was 12.73°. With vertically polarised light at 532nm it is possible to use a simple minimum deviation approach to measure n_y and the value so obtained was 1.7230. With horizontally polarised light and a series of different incidence angles it is possible to obtain a relatively accurate value for n_x (2.17), since the dielectric x-axis is nearly parallel with $(10\bar{2})$. It follows that n_z will not be accurately measurable, requiring a prism cut in the orthogonal direction. Accordingly, Maker fringe spacings were used to calculate n_x and n_y for 1060nm light. It can be shown that to the order of $\sin^2\theta$ a plot of M (fringe number) against $\sin^2\theta$ has a slope of L/λ. $[f(2\omega)-f(\omega)]$. Here L is the sample thickness and λ is the wavelength of the fundamental. For ordinary rays, functions, f, are reciprocals of the relevant refractive indices, whereas for extraordinary rays the functions are;

[n(along the intersection of the plane of incidence and the face of the sample)] / [n(along the normal to the face)]^2.

From two such plots, which were very linear for incidence angles up to $\sim 40^\circ$ n_x and n_y at 1064nm were calculated. There is some ambiguity of sign in these difference formulae. Generally it is assumed that the refractive index at 532nm will be larger than the value at 1064nm along the same dielectric axis. Additionally, the prism was used to obtain refractive index data at a variety of visible wavelengths and a Lorentz equation was used to extrapolate these to 1064nm. The resulting value for n_x at 1064nm was 1.9401 compared with the Maker fringe value of 1.899. For n_y the extrapolated value of 1.6525 compared well with the Maker fringe value of 1.643. This latter was the higher of two values arising from the sign ambiguity. n_z at 532 and 1064nm could be obtained from solution of two simultaneous equations for further Maker fringe curves. However, these signals were of very low amplitude and the solutions of the equations depended on very small differences. Accordingly, n_z (532nm) was deduced from the type II phase matching (Fig. 4) and the n_z for 1064nm calculated from one of the Maker fringe relations. The remaining equation was used as a check for the consistency of the results and predicted a slope of the e-ray type plot (corrected for L/λ) of 0.0465. This compared with a value of 0.0413 deduced from the n_z values. The resulting refractive index values are shown in table 1.

Table 1

	X	Y	Z
n^{1060nm}	1.90	1.643	1.7911
n^{532nm}	2.17	1.7230	1.8318

Table 2 shows the results of d_{ij} calculations. These values are referenced to the dielectric axes and are relative to d_{11} of quartz. The SHG intensities relative to quartz (I_{532nm}) are also given, together with the coherence length (l_c).

Table 2

	I_{532nm}	d_{ij}	$l_c/\mu m$
d_{11}	80.5	311	0.98
d_{33}	0.5	3	6.5

No dispersion of the dielectric axes was observed by examination under the polarising microscope ($\pm$ 2 deg) between 440 and 630nm. The Z dielectric axis is ~ 18° from the c crystallographic axis in an anticlockwise direction about b observing along b.

For rotation about the c-axis, phase matching signals are seen for the (100) face at -8° and $+4^{\circ}$ from normal incidence. However, the classification of these as types I or II is complicated by the non-coincidence of the c-axis with any dielectric axis.

The phase matching locus was calculated from the above refractive index data. This is shown in Fig. 5, referred to the dielectric axes. Phase matching coincides closely with a^{*} (piezoelectric axis).

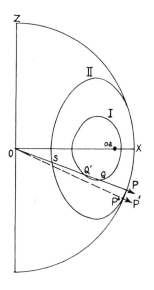

Fig. 5. Stereographic projection (along b-axis) of the phase matching locus of NMBA. I and II denote types. X and Z represent the dielectric axes. oa is one 1064nm optical axis.

For rotation of the (100) face about the Y-dielectric axis, propagation directions are represented by points on the circumference of the stereogram. The line OP represents the direction of propagation in the laboratory frame of reference. Clearly only type II phase matching will be seen and it will appear at near normal incidence to the (100) face. This agrees well with Fig. 4.

For rotation about the Z-dielectric axis, the angle of refraction varies over OP with P again the normal incidence position. S and Q' are at inaccessibly large angles of incidence and Q is also just at the limit of our scan angle and thus not seen. If the orientation of the Z rotation axis is very slightly misaligned, or the refractive indices are slightly in error, two peaks will be observed close on either side of normal incidence as represented by P" and its conjugate below the ZOX plane. The type I locus will be bypassed. This also agrees well with the experimental observations. Further work is required on this phase matching locus and the refractive index values. This will use data from the (001) face currently under preparation. Further d_{ij} values will be deduced and these, of course, determine the intensity of the phase matched signal along the above locus.

ACKNOWLEDGEMENTS

We wish to thank the SERC and the DTI for financial support under the Joint Opto-Electronic Research Scheme and the former for purchase of the Fourier transform spectrometer. We are also grateful for the help of Dr. D.I. Bishop (BDH Ltd) in provision of material, Prof. M.B. Hursthouse and Dr A.S. Karaulov for X-ray structural information, and Dr. T. Maclean for discussions on crystal growth.

REFERENCES

1. D.S. Chemla and J. Zyss (Eds.). "Non-Linear Optical Properties of Organic Molecules and Crystals", 1 and 2, Academic Press, N.Y., (1987).

2. R.T. Bailey, S. Blaney, F.R. Cruickshank, S.M.G. Guthrie, D. Pugh and J.N. Sherwood. J. Appl. Phys. B 47, 83 (1988).

3. B.J. McArdle and J.N. Sherwood. "Advanced Crystal Growth", (Ed. P.M. Dryburgh, B. Cockayne and K.G. Barraclough, p179, (1987).

4. O.S. Filipenko, V.D. Shigorin, V.I. Ponomarev, L.O. Atovmyan, Z.Sh. Safina and B.L. Tarnopol'skii, Sov. Phys. Crystallography, 22, 305, (1977).

5. J.N. Sherwood in preparation.

6. R.T. Bailey, F.R. Cruickshank, S.M.G. Guthrie, B.J. McArdle, H. Morrison, D. Pugh, E.E.A. Shepherd, J.N. Sherwood, G.S. Simpson and C.S. Yoon. Mol. Cryst. Liq. Cryst. 166, 267, (1989).

7. S.N. Black, R.J. Davey and T. McLean. Mol. Cryst. Liq. Cryst. In press.

8. P. Hartman and P. Bennema. J. Crystal Growth, 49, 157, (1980).

Inst. Phys. Conf. Ser. No 103: Section 2.1
Paper presented at Int. Conf. Materials for Non-linear and Electro-optics, Cambridge, 1989

Nonlinear optical and electro-optic properties of 2-(N-prolinol)-5-nitropyridine (PNP) crystals

K. Sutter, Ch. Bosshard, L. Baraldi and P. Günter

Institute of Quantum Electronics, ETH Hönggerberg, CH-8093 Zürich, Switzerland

ABSTRACT: We have investigated the nonlinear optical and electro-optic properties of 2-(*N*-prolinol)-5-nitropyridine (PNP). Three second-order nonlinear optical coefficients were measured at fundamental wavelengths of 1064 and 1318 nm. The phase matching conditions for angle tuned second harmonic generation at these wavelengths were calculated from the refractive index data and verified experimentally. Maximum second harmonic generation efficiencies of up to 20 % were reached in a crystal of a thickness of 1.2 mm at a fundamental peak intensity of 10 MW/cm^2. Two electro-optic coefficients were determined at 514 nm and 633 nm.

1. INTRODUCTION

Organic donor-acceptor systems are known to show very high optical nonlinearities (Nicoud *et al* 1987). One of these materials is 2-(N-prolinol)-5-nitropyridine (PNP) (Fig. 1). It was first proposed as a potentially interesting nonlinear optical crystal by Twieg *et al* (1983 and 1986). We have grown single crystals of PNP from methanol-water solution using a temperature lowering technique. The detailed crystal growth procedure will be published elsewhere (Wang *et al* 1989). Single crystals having volumes of up to 500 mm^3 were obtained. On these samples linear and nonlinear optical measurements were carried out.

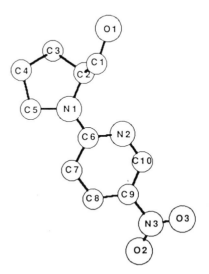

Fig. 1. A single molecule of 2-(N-prolinol)-5-nitropyridine (PNP).

2. REFRACTIVE INDICES AND ABSORPTION

PNP crystals belong to the monoclinic point group 2 (Twieg *et al* 1986). The projection of a unit cell along the crystallographic b-axis is shown in Fig. 2. For point group symmetry 2 one axis of the index ellipsoid is fixed parallel to the crystallographic symmetry axis b. The other two axes lie in the crystallographic ac-plane.

All optical data will be given with respect to the dielectric system x,y,z which is defined by the main axes of the index ellipsoid. Our measurements have shown that one of these axes is nearly parallel to the crystallographic direction [1,0,-1]. We define this axis as x. y is parallel to the crystallographic b-axis (Fig. 2). The angle υ between the dielectric direction x and the crystallographic direction [1,0,-1] is small ($\upsilon \leq 2°$) and changes less than 0.5° in the spectral range of 490 to 680 nm.

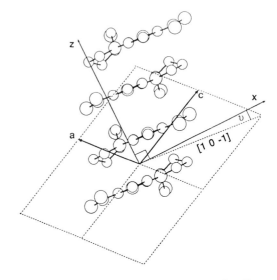

Fig. 2. Projection of the molecules of two unit cells along the b axis. The dielectric axes x and z are shown.

The refractive indices were determined with an interferometric method (Sutter *at al* 1988b) and have been published (Sutter *et al* 1988a). The resulting dispersion of the principal refractive indices are shown in Fig. 3.

The optical absorption was measured on a plate cut perpendicularly to the b-axis (Sutter *et al* 1988a). The transmission cut-off for short wavelengths lies at 466 nm for light polarized along z and at 490 nm for light polarized along x.

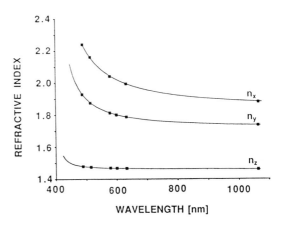

Fig. 3. The dispersion of the refractive indices of PNP at room temperature.

3. NONLINEAR OPTICAL PROPERTIES

For crystals of point group symmetry 2 the nonlinear optical polarization $P(2\omega)$ induced by a field $E(\omega)$ is given by

$$
\begin{aligned}
P_x(2\omega) &= 2\varepsilon_0 \left(d_{xyz}E_yE_z + d_{xxy}E_xE_y\right) \\
P_y(2\omega) &= \varepsilon_0 \left(d_{yxx}E_x^2 + d_{yyy}E_y^2 + d_{yzz}E_z^2 + 2d_{yxz}E_xE_z\right) \\
P_z(2\omega) &= 2\varepsilon_0 \left(d_{zyz}E_yE_z + d_{zxy}E_xE_y\right)
\end{aligned}
$$

To simplify the interpretation of the results the dispersive rotation of the axes x and z around y was neglected for all our calculations.

From the molecular arrangement it can be shown (Twieg *et al* 1986) that d_{yxx}, d_{yyy} and d_{xxy} should be dominant while all other coefficients are close to zero.

We have measured the coefficients d_{yxx}, d_{yyy} and d_{xxy} by the Maker-fringe method. In our experimental set-up we used a flashlight pumped Q-switched Nd:YAG laser operating at wavelengths of 1064 or 1318 nm. Table I shows the results of our experiments, measured with respect to quartz with $d_{11}^{\text{quartz}} = 0.4$ pm/V:

Table I: The nonlinear optical coefficients of PNP:

Wavelength [nm]	d_{yxx} [pm/V]	d_{yyy} [pm/V]	d_{xxy} [pm/V]
1064 nm	67.9 ± 8	21.6 ± 4	53.4 ± 6
1318 nm	38.9 ± 5	15.3 ± 3	35.6 ± 5

Some of the coefficients at 1064 nm (d_{yxx} and d_{yyy}) are somewhat larger than a preliminary result reported earlier (Sutter *et al* 1988a). This difference is due to increased crystal quality and better contrast of the Maker-fringe curves.

In the one-dimensional molecular model the nonlinear optical susceptibilities for frequency doubling can be expressed as follows (Zyss *et al* 1987):

$$
d_{IJK} = N f_I(2\omega) f_J(\omega) f_K(\omega) \, b_{IJK}
$$

$$
\text{with} \quad b_{IJK} = \frac{1}{N_g} \sum_{s=1}^{N_g} \cos \theta_{Iz}^{(s)} \cos \theta_{Jz}^{(s)} \cos \theta_{Kz}^{(s)} \, \beta_{zzz}
$$

$$
\text{and} \quad \beta_{zzz} = \frac{3\hbar^2}{m_e} \frac{W}{\left(W - \hbar^2\omega^2\right)\left(W - 4\hbar^2\omega^2\right)} f\delta
$$

where N is the number of molecules per volume, $f_L(\omega)$ are the local field factors, N_g is the number of equivalent sites in the unit cell, θ_{Lz} is the angle between the direction L and the molecular CT-axis, β_{zzz} is the (one-dimensional) molecular hyperpolarizability, W the oscillator energy, f the oscillator strength and δ the change of molecular dipole moment between ground-state and CT-state. The oscillator energy or the oscillator frequency $\omega_0 = W/\hbar$ can be obtained by fitting the refractive indices by a one-oscillator formula. For PNP we have $\omega_0 = (4.804 \pm 0.25) \cdot 10^{15}$ s^{-1} (Sutter *et al*, 1988a).

From these formulas the dispersion of the nonlinear optical susceptibility can be estimated. Using the (rather crude) Lorentz-approximation for the local field factors

$$f_L(\omega) = \frac{n_L^2(\omega) + 2}{3}$$

we have calculated the ratios of the d-coefficients at 1318 nm at 1064 nm. The comparison of theoretical and experimental results are given in Table II:

Table II: Ratios of nonlinear optical susceptibilities at the wavelengths 1064 nm and 1318 nm. Comparison between experimental and theoretical results.

	$\dfrac{d_{yxx}(1064nm)}{d_{yxx}(1318nm)}$	$\dfrac{d_{yyy}(1064nm)}{d_{yyy}(1318nm)}$	$\dfrac{d_{xxy}(1064nm)}{d_{xxy}(1318nm)}$
theoretical	1.61 ± 0.1	1.58 ± 0.1	1.68 ± 0.1
experimental	1.75 ± 0.3	1.41 ± 0.3	1.50 ± 0.3

Even though some rough approximations were used the theoretical and experimental values are in good agreement.

4. PHASE MATCHING EXPERIMENTS

Angle tuned phase matching for second harmonic generation can be found in PNP at both fundamental wavelengths of the Nd:YAG laser, 1064 nm and 1318 nm. We have performed phase matching experiments on a plate cut perpendicularly to the dielectric z-axis. The plate was mounted on a rotation stage which allowed a rotation of the sample around the y-axis by an angle ϕ and subsequent tilting around the x-axis by θ. To increase the internal angles of the beams the sample was placed in an immersion fluid with a refractive index of 1.440. The phase matching angles were determined experimentally and calculated from the refractive indices. For the calculation of the refractive indices the dispersion formula (Sutter *et al* 1988a) was used. The results are shown in Fig. 4.

The agreement between the calculated phase matching areas (dotted areas) and our experimental points is a indication for the accuracy of the refractive index data.

The efficiency for type I phase matched second harmonic generation was measured on a 1.2 mm thick sample cut perpendicular to z, rotated by $\phi = 22.9°$ around the y-axis ($\theta = 0°$). No immersion fluid was used in this experiment. For a crystal of a thickness of 1.2 mm and a fundamental peak intensity of 10 MW/cm^2 (beam waist = 600 μm) a maximum efficiency of 20% was reached. We could not further increase the efficiency by increasing the fundamental intensity up to 30 MW/cm^2. This deviation from normal behaviour might be explained by thermal effects.

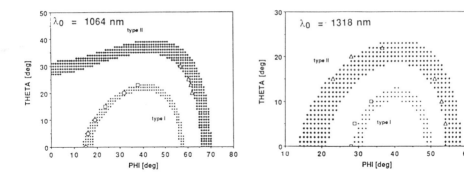

Fig. 4. Phase matching loci for a PNP plate cut perpendicular to z in immersion fluid (n = 1.440). The dots represent calculated phase matching regions. Squares and triangles stand for experimental points. For a definition of rotation and tilt angles ϕ and θ see text. For the left graph the fundamental wavelength is 1064 nm, for the right graph 1318 nm.

Theoretically the power conversion efficiency $\eta = P^{2\omega}/P^\omega$ for phase matched SHG neglecting pump depletion is given by (Yariv 1975)

$$\eta = K \frac{2\omega^2}{\varepsilon_0 c^3} \frac{d_{eff}^2}{n^3} L^2 I_\omega$$

where ω is the angular fundamental frequency, d_{eff} the effective nonlinear optical coefficient, n the refractive indices (= $n_y(2\omega)$ in our configuration), L the crystal length and I_ω the fundamental intensity. The factor K is determined by the temporal pulse shape. For continuous light we have K=1, for Gaussian pulses K = $1/\sqrt{2}$. From our experimental data (η = 20%, I_ω = 10 MW/cm^2) we obtain d_{eff} = 22 pm/V.

d_{eff} can also be calculated from the nonlinear optical coefficients by using the formula

$$d_{eff} = \cos^2\gamma \, d_{21} + \sin^2\gamma \, d_{23} + 2 \cos\gamma\sin\gamma \, d_{25}$$

where γ is the angle between the electric field and the x-axis as calculated by Snell's law and the dielectric tensor (γ = 17.7° in our configuration). So we have d_{eff} = 0.89 d_{21} + 0.11 d_{23} + 0.63 d_{25} ≈ 60 pm/V (assuming d_{23} ≈ 0 and d_{25} ≈ 0). This is much higher than the value obtained from the phase matching experiment. This discrepancy confirms the assumption that the conversion efficiency is reduced for high intensity, e.g. by thermal effects.

5. ELECTRO-OPTIC MEASUREMENTS

The electro-optic coefficients r_{ijk} are defined by the change of the inverse dielectric tensor ε^{-1} under an applied electric field **E**:

$$\Delta(\varepsilon^{-1})_{ij} = \sum_k r_{ijk} E_k$$

PNP, belonging to point group symmetry 2, has 8 non-zero electro-optic coefficients: r_{12}, r_{22}, r_{32}, r_{41}, r_{43}, r_{52}, r_{61} and r_{63}.

So far we have only carried out measurements on one crystal. It was polished perpendicular to the z-axis and the electric field was applied along the y-direction. Thus, by using a Michelson-interferometer the coefficients r_{12} and r_{22} could be determined. The experimental procedure is described by Sutter *et al* (1988b). The results at an optical wavelength $\lambda = 633$ nm were $r_{12} = (13.1 \pm 2)$ pm/V and $r_{22} = (13.1 \pm 2)$ pm/V, at $\lambda = 514$ nm we obtained $r_{12} = (20.2 \pm 3)$ pm/V and $r_{22} = (28.3 \pm 4)$ pm/V. Changing the frequency of the applied electric field between 1 kHz and 100 kHz had no influence on the electro-optic coefficients. The applied fields were in the order of 100 V/cm. The crystal showed no sign of degradation or fatigue over a period of several weeks of measurements.

ACKNOWLEDGEMENTS

The authors are grateful to J. Hajfler for his expert sample preparation. This work has been supported by the Swiss National Science Foundation (NFP 19: *Materials for Future Technology*).

REFERENCES

Nicoud J F and Twieg R J 1987 *Nonlinear Optical Properties of Organic Molecules and Crystals* ed D S Chemla and J Zyss (Orlando: Academic Press), Vol. 2, pp. 221-254
Sutter K, Bosshard Ch, Wang W S, Surmely G and Günter P 1988a *Appl. Phys. Lett.* **53**(19) 1779
Sutter K, Bosshard Ch, Ehrensperger M, Günter P and Twieg RJ 1988b *IEEE J. Quantum Electron.* **24**(12) 2362
Twieg R J and Dirk C W 1986 *J.Chem.Phys.* **85** 3537
Twieg R J and Jain K 1983 *Nonlinear Optical Properties of Organic and Polymeric Materials* ed D Wiliams (Washington, D.C.: Am. Chem. Soc.)
Wang W S *et al* 1989, to be published
Yariv A 1975 *Quantum Electronics* (John Wiley and Sons, New York)
Zyss J and Chemla D S 1987 *Nonlinear Optical Properties of Organic Molecules and Crystals* ed D S Chemla and J Zyss (Orlando: Academic Press), Vol. 2, pp. 23-192

Inst. Phys. Conf. Ser. No 103: Section 2.1
Paper presented at Int. Conf. Materials for Non-linear and Electro-optics, Cambridge, 1989

Optical and nonlinear optical properties of 4-(N,N-dimethylamino)-3-acetamidonitrobenzene single crystals

P Kerkoc, M Zgonik*, K Sutter, Ch Bosshard and P Günter
Institute of Quantum Electronics, Swiss Federal Institute of Technology
ETH -Hönggerberg, CH-8093 Zürich, Switzerland

ABSTRACT: A method of growing high quality 4-(N,N-dimethylamino)-3-acetamidonitrobenzene (DAN) single crystals is presented. Crystal plates with dimensions of up to 10x5x2 mm³ were produced and optically characterized. Large birefringence and strong dispersion of the refractive indices were measured allowing two branches of type I and two of type II phase-matched second-harmonic generation. From effective second-order nonlinear optical susceptibilities measured along the branches all nonlinear susceptibilities were evaluated, the highest being $d_{23} = (50 \pm 15)$ pm/V.

1. Introduction

Molecular crystals with charge correlated and highly delocalized π-electron states such as nitroanilines appeared very attractive in recent years, since very large nonlinear optical susceptibilities were measured in some of these materials (Williams 1983, Sutter *et al* 1988 and Bosshard *et al* 1989). One of them, 4-(N,N-dimethylamino)-3-acetamidonitrobenzene (DAN) was first proposed by Twieg *et al* (1983) as a potentially efficient nonlinear crystal. Single crystals of DAN were grown either by slow cooling of a saturated methanol solution by Norman *et al* (1987) or from a melt by the Bridgman method by Baumert *et al* (1987), and Bailey *et al* (1988). The crystal structure of DAN determined by Twieg *et al* (1986) is monoclinic with space group $P2_1$ and cell parameters $Z = 2$, $a = 0.4786$ nm, $b = 1.3053$ nm, $c = 0.8736$ nm and $\beta = 94.43°$.

Efficiencies up to 20% were reached with type I phase-matched frequency doubling for the wavelength of $\lambda = 1064$ nm (Norman *et al* 1987). The first measurement of an effective quadratic nonlinear optical (NLO) susceptibility gave a value of (27 ± 3) pm/V (Baumert *et al* 1987). In this work we report on the preparation of DAN single crystals, and fully characterize their optical and nonlinear optical properties.

2. Growth of DAN single crystals

Dimethyl sulfoxide as solvent was saturated with DAN at a temperature of 32°C and crystals were grown from excess feed material in a temperature difference procedure described by Arend *et al* (1986). The method is based on thermal convection with spontaneous nucleation in the cold part of the system. Within two weeks high optical quality DAN crystal plates with dimensions of up to 10x5x2 mm³ were grown.

---

*Permanent address: J. Stefan Institute, E. Kardelj University, Jamova 39,
61111 Ljubljana, Yugoslavia

3. Refractive indices measurements

A sample with surfaces normal to the y-axis was observed under polarizing microscope to determine the orientation of the principal axes x and z of index ellipsoid within the sample. The angle ϕ between the normal to an as grown crystal plate (001) and the x-axis is $54.9° \pm 0.1°$ as shown in Figure 1. The orientation of the indicatrix showed no dispersion for light in the spectral range from 488 to 1064 nm.

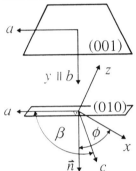

Fig.1. The largest surface is parallel to the (001) crystallographic plane. Drawing shows orientation of the principal axes of the indicatrix (x,y,z). The angle ϕ between normal to the plate n and the x-axis is $54.9°\pm0.1°$. The angle β between the crystallographic axes a and c is $94.43°$.

For measuring the refractive indices an interferometric method of Shumate (1966) was attempted. The method uses a Michelson interferometer with a rotating crystal plate in one arm of it. A platelet oriented parallel to (001) crystallographic plane was used for the determination of the refractive index $n_y = n_b$. Different wavelengths of Ar$^+$, He-Ne, dye (rhodamine 6G) and Nd:YAG laser were used in the measurements and the results are presented in Table I and Figure 2. n_y was also measured by an immersion fluid method and the results agree well as seen in Figure 2.

A new procedure had to be developed for the interferometric measurement of n_x and n_z refractive indices since only (001) samples could be polished. The crystal plate was rotated around the y-axis on the stage. The incident laser beam was polarized at 45° to the y-axis to excite both eigen-polarizations one parallel to y and the other lying in the $x - z$ plane. The outgoing beam interfered at the detector after traversing an analyzer also inclined by 45° to the y-axis. The starting orientation of the sample was always chosen such that the beam propagated along the optic axis - which lies almost perpendicularly to the (001) plane - and the optical path difference was zero. The sample with thickness d was then rotated clockwise and counter-clockwise and the detector output was recorded. The optical path length difference at an external incidence angle θ measured from the normal incidence is given by:

$$m\lambda = d\left(\sqrt{n_{xz}^2 - \sin^2\theta} - \sqrt{n_y^2 - \sin^2\theta} \right), \tag{1}$$

and could be absolutely determined from the number of interference fringes m counted from the starting orientation where $m = 0$. The refractive indices n_{xz} were determined from (1) at several incidence angles. The internal propagation angles ϑ were calculated from Snell's law. The principal refractive indices n_x and n_z were then calculated from

$$n_{xz}^{-2}(\vartheta) = n_x^{-2} \sin^2(\phi + \vartheta) + n_z^{-2} \cos^2(\phi + \vartheta) \ . \tag{2}$$

Because the light propagation direction along the z -axis can be closely approached in our samples, the index n_x was determined with higher accuracy than n_z. The procedure was repeated with several laser wavelengths and the results are presented in Table I.

TABLE I. Refractive indices of solution grown DAN crystal at room temperature.

λ (nm)	n_x (± 0.007)	n_y (± 0.005)	n_z (± 0.009)
496.5	1.574	1.779	2.243
514.5	1.557	1.748	2.165
532	1.554	1.732	2.107
585	1.545	1.701	2.005
632.8	1.539	1.682	1.949
1064	1.517	1.636	1.843

The dispersion data were fitted with a one oscillator model:

$$n^2 - 1 = \frac{q}{\lambda_0^{-2} - \lambda^{-2}} + A \ , \tag{3}$$

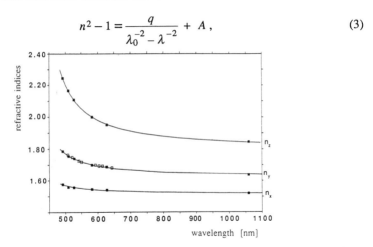

Fig. 2. Dispersion of the refractive indices of DAN at room temperature. Points □: measured by the immersion method. Points ■:interferometrically measured values.

where $q = p/(2\pi c)^2$, p is the oscillator strength and λ_0 is the wavelength of the main oscillator. A is a constant contribution from all other oscillators not considered in this simple model. The refractive index data and the corresponding fit are presented in Figure 1. The fitting parameters q, λ_0 and A are given in Table II.

TABLE II. Fitting parameters of the dispersion relation (3) for the three principal refractive indices.

	q (10^{12} m^{-2})	λ_0 (10^{-6} m)	A
n_x	1.0879	0.3681	1.1390
n_y	1.9858	0.3933	1.3290
n_z	4.0908	0.4194	1.5379

¹ Nonlinear optical susceptibility measurements

In measuring the second-order susceptibility tensor $\underline{d}$ of DAN we assumed symmetry upon interchanging all three indices. This is known as Kleinman's symmetry. $\underline{d}$ is then characterized by four different coefficients: d_{21}, d_{22}, d_{23}, and d_{25}. Since only (001) plates could be prepared, the Maker-fringe method (Kurtz 1975) could only be used for the measurement of d_{22} and gave a value of 5.2 pm/V with an uncertainty of 20 %. The relatively large birefringence provides a wealth of phase-matching directions for second-harmonic generation (SHG). Directions for angle-tuned collinear type I and type II phase matching for the fundamental wavelength of 1064 nm were calculated from our data and are presented in Figure 2. For the measurements of the phase-matching properties 0.2 mm to 0.3 mm thick (001) crystal plates were used, which were not polished but were sandwiched between glass plates by silicone rubber glue to reduce scattering and to allow the use of an index matching fluid. Index matching was necessary to reach larger internal angles of light propagation. The sample was mounted on a goniometer stage and immersed into a cuvette containing the liquid with the index of refraction of 1.433. The laser beam was weakly focused to a spot size of 0.5 mm Gaussian beam (FWHM of the beam) to avoid problems with beam walk–off. The spread of propagation directions for type II phase matching was not taken into account since the resulting error is within the overall accuracy. Figure 3 shows good correspondence between the predicted and measured directions of phase matching.

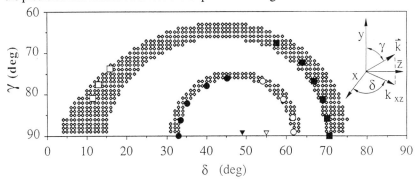

Fig.3. Phase-matching configurations of DAN crystals. Calculated points are marked with ◊. The outer region allow type II and the inner one type I phase matching. A mismatch of 0.007 in the calculated value of the n^{2w} was allowed. Experimental points ■, O, ● and □ correspond to different combinations of incident and generated light polarizations as described in Table III. Point ▼ marks the optical axis and ▽ the surface normal. The coordinate system is defined in the figure with $\vec{k}$ and k_{xz} being the the propagation direction and its projection to x - z plane.

For each one of the experimental points in Figure 3 the effective nonlinear coefficient d_{eff} was measured. Calibration of the apparatus was done by a comparison with a measurement of d_{33} = 27.4 pm/V of $KNbO_3$ (Baumert *et al* 1984). The overall accuracy of the measurements was better than 20 %. The highest effective nonlinear coefficient with a value of 27 pm/V was measured in the x - z plane in branch b) in agreement with the previous data of Baumert et al (1987). For this configuration the beam walk-off between fundamental and second harmonic wave is 10.8°. In other branches d_{eff} are smaller. Every measured d_{eff} was expressed as a linear combination of d_{21}, d_{22}, d_{23}, and d_{25}. Individual coefficients were then obtained through multiple linear regression giving: d_{21} = (1.5 ± 2) pm/V, d_{23} = (50 ± 15) pm/V and d_{25} = (1.5 ± 2) pm/V (Kerkoc *at al* 1989). The prevalence of d_{23} over d_{21} and d_{25} is the reason for the large inaccuracy of the latter two coefficients.

5. Second harmonics generation (SHG) conversion efficiency and damage resistance

We have performed frequency doubling experiments with high conversion efficiency of up to 15%. The sample was a 1.18 mm thick (00$\bar{1}$) plate oriented for type I phase-matching in the x-z plane. Two Nd:YAG lasers operating at 1064 nm were used in the measurements.

The first was a CW pumped mode-locked and Q-switched laser with 0.4 MW peak power. When focused to a beam waist diameter of 0.3 mm the highest measured SHG conversion efficiency was 8%. Tighter focusing does not increase the efficiency due to the large walk-off between infrared and green beams which is around 10.8°. The peak power density 500 MW/cm^2 used in this experiment is rather high, but is well below damage threshold which was estimated to be more than 5 GW/cm^2 for Q-switched and mode-locked pulses of 100 ps, so higher efficiencies are possible with high power lasers as already shown in relatively thin crystals. SHG efficiencies and damage threshold were then investigated using a Q-switched flash-lamp pumped laser with 15 ns pulse width and repetition rate 30 Hz. Damage threshold for these longer pulses was found to be much lower and DAN samples showed first irreversible defects at a power density of 80 MW/cm^2. However a conversion efficiency of 15 % was obtained with a Gaussian beam with waist of 0.6 mm and peak power density of 40 MW/cm^2.

In the range of 10 to 50 MW/cm^2 of peak power density the conversion efficiency was found to be nearly independent of input intensity a result which was reported previously (Norman *et al* 1987). This anomalous behavior of SHG and also relatively low conversion efficiency obtained with the first laser may be due to inhomogeneous heating of the crystal both by slightly absorbed fundamental and second-harmonic waves. From the external acceptance angle and shift of the phase-matching direction by temperature tuning, we calculate a temperature acceptance angle of 16 °C for a 0.357 mm thick sample. Heating of the crystal of up to 10-20 °C by laser radiation could therefore cause a considerable change in phase-matching conditions and decrease the SHG conversion efficiency. Further experiments are necessary to understand SHG at higher power levels in DAN.

6. Conclusion

The large optical anisotropy of DAN allows several phase-matched configurations for nonlinear processes. The phase-matchable susceptibility d_{23} of around 50 pm/V is one of the largest found so far, comparable to $d_{21} \approx 50$ pm/V of organic 2-(N-prolinol)-5-nitropyridine (PNP) measured by Sutter *et al* (1988). The Figure of merit of DAN $d_{eff}^2/n^3 = 140$ [pm/V]^2 is more than 10 times larger than that of potassium titanyl phosphate (KTP) Belt *et al* (1985). The optical quality of the crystals produced was very good. It can also be judged from the regularity of the angular dependence of the SHG efficiency curves. Both angle tuned phase-matching and Maker fringe patterns showed high quality of the crystals with no traces of twins. Another advantage of DAN with respect to optical applications is its large damage threshold which was estimated to be higher than 5 GW/cm^2 for pulses of 100 ps width. Cutting and polishing of DAN crystals still presents a problem as only (001) faces could be finished to high optical quality. New procedures will have to be developed for the preparation of differently cut crystals. Its properties make DAN a promising material for future applications in the field of parametric oscillators and frequency converters.

7. Acknowledgement

The authors are grateful to J. Hajfler for his expert sample preparation. This work has been supported in part by the Swiss National Science Foundation (NFP 19): Materials for Future Technology.

8. References

Williams D J ed. *Nonlinear Optical Properties of Organic and Polymeric Materials,* Am. Chem. Soc. Symp. Ser. 233, (American Chemical Society, Washington, D.C.,1983)

Sutter K, Bosshard Ch, Ehrensperger M, Günter P, and Twieg R J 1988 IEEE J. Quantum Electron. **24** 2362 .

Bosshard Ch, Sutter K,Günter P, and Chapuis G 1989 J. Opt. Soc. Am. B 6, 721

Twieg R J and Jain K, in *Nonlinear Optical Properties of Organic and Polymeric Materials,* Am. Chem. Soc. Symp. Ser. 233, ed. by D.J. Williams (American Chemical Society, Washington, D.C.,1983) pp 57-80

Norman P A, Bloor D, Obhi J S, Karaulov S A, Hursthouse M B, Kolinsky P V, Jones R J and Hall S.R.1987 J. Opt. Soc. Am. **B 4** 1013

Bailey R T, Cruickshank F R, Guthrie S M G, McArdle B J, McGillivray G W, Pugh D, Shepherd E E A, Sherwood J N, Simpson G S and Yoon C S Sept. 1988 to be published in *Proc. Int. Con. Opt. Sci.& Eng. (ECO)* Hamburg

Twieg R J and Dirk C W 1986 Research Report RJ 5329 (54799), IBM Almaden Research Center San Jose CA 95120-6099.

Baumert J-C, Twieg R J, Bjorklund G C, Logan J A and Dirk C W 1987 Appl. Phys. Lett. **51** 1484

Arend H, Perret R, Wüest H and Kerkoc P 1986 J. of Cryst. Growth **74** 321

Shumate M S 1966. Appl. Optics **5** 327

Kurtz S K, in *Quantum Electronics*, ed.Rabin H and Tang C L (Academics New York 1975) p 209

Kerkoc P, Zgonik M, Sutter K, Bosshard Ch and Günter P Appl. Phys. Lett. (to be published)

Sutter K, Bosshard Ch, Wang W S, Sumerly G and Günter P 1988 Appl. Phys. Lett. **53** 1779

Belt R F, Gashurov G and Liu Y S 1985 *Laser Focus* **21** 110

Inst. Phys. Conf. Ser. No 103: Section 2.2
Paper presented at Int. Conf. Materials for Non-linear and Electro-optics, Cambridge, 1989

Characterization of organic nonlinear materials

G R Meredith, L T Cheng, H Hsiung, H A Vanherzeele, F C Zumsteg

E I Du Pont De Nemours & Co, Central Research & Development, Wilmington, Delaware 19880-0356, USA

ABSTRACT: Nonlinear optical characterization of molecular materials is discussed. Examples include studies of trends in hyperpolarizabilities as determined in solutions, characterization of novel molecular crystals and studies of molecular orientation in thin films.

1. INTRODUCTION

The molecular nature of organic materials allows a particular view of optical properties which is functionally unique compared with inorganics and semiconductors. If attention is restricted to nonresonant situations, the polarization phenomena which govern linear and nonlinear optical processes can be considered to be the result of the projection of local microscopic, or molecular, polarization processes to the macroscopic level. This model can be used in at least three ways:

- one may empirically study convenient ensembles for the purpose of determining molecular properties, which should ideally in this picture be transferable from ensemble to ensemble,
- one may use the nonlinear optical response as a probe of configuration and its changes within an ensemble, or
- one may prepare and assemble molecular units to create nonlinear optical media with desirable properties.

The large magnitudes of nonlinear polarizabilities, the diversity of molecular and ensemble structures achievable, the new avenues of fabrication, as well as the new science which can be uncovered have been motivating factors for research on nonlinearity of organics (Meredith 1988). In this work the molecular picture has been implicit, though naturally questionable. Herein we will not probe the validity of the model, rather illustrate activities of the three types listed above. By the use of carefully designed experimentation it is possible to uncover new trends and to evaluate the successes of materials strategies intended to provide new nonlinear media. The examples presented below include studies of molecular electronic hyperpolarizabilities and of "configuration/alignment" in crystals and thin films. Except in the molecular studies, where it will be shown to be essential to include the third-order nonlinearity, the discussion pertains to the development of second-order nonlinear optical media.

2. OPTICAL CHARACTERIZATION

Nonlinear optical processes are diverse, as are the mechanisms which govern a material's participation in them. For processes which involve minor perturbations of the material and where the material response times are faster than the optical pulse envelope time, it is common to use polarization expansions in powers of the frequency components of the electric field, focusing attention on the portion of induced polarization at the frequency of the observed light wave,

$$P^\omega 0 = \chi^{(1)}(-\omega_0;\omega_0)\cdot E^\omega 0 + \Sigma_{ij}\,\chi^{(2)}(-\omega_0;\omega_i,\omega_j)\cdot\cdot E^\omega iE^\omega j +$$
$$\Sigma_{ijk}\,\chi^{(3)}(-\omega_0;\omega_i,\omega_j,\omega_k)\cdot\cdot\cdot E^\omega iE^\omega jE^\omega k + \dots ,$$

with a similar expansion of induced molecular polarization in terms of the tensorial linear polarizability α and hyperpolarizabilities β, γ, etc. (One must count all contributions and routes to this frequency, since the optical fields and polarizations which can generate or be generated by them are not restricted to the power or "order" of immediate interest to the experimentalist. This applies to both Maxwellian and "local" fields (Meredith 1982a and b).)

In an earlier paper nonlinear optical processes were classified as Types A, B and C for the purpose of describing common features (Meredith et al 1989b). The reader is referred there for details omitted herein. Processes where optical retardation variations are induced are easiest to calibrate and are often simple to use since experimental artifacts can usually be simply avoided. In contrast, when light at new frequencies is generated, complications arise due to mismatch of phase and/or energy propagation between waves resulting in nonintuitive behavior and sensitively variable coherent interferences of fields at detectors. However, the occurrence of coherent interferences, both within materials' susceptibilities and between media placed along the optical path, allow useful secondary calibrations and can be exploited, as in the solution techniques described below, to improve the precision of a technique through destructive interferences. Two experimentally complicating factors are the tensorial and frequency dispersive nature of the material susceptibilities and molecular hyperpolarizabilities, it being fair to say that for no material has a nonlinear tensor been characterized fully with respect to anisotropy and dispersion. Thus care must be taken to consider the quantities actually reported and the differences which may exist between them when materials are compared (Meredith et al 1983a). Finally, it is emphasized that with condensed phase materials the susceptibilities can be quantified to high degrees of precision. However, relating these to microscopic or molecular properties relies on models of dielectric behavior. These models are often simple, but not necessarily accurate (Meredith 1987). One finds, then, that different analyses of the same experimental data can yield molecular hyperpolarizabilities which differ by large amounts (e.g. by 50 % or more), however, consistent analysis methodology among sets of experimental data can reliably show trends (Cheng et al 1989).

As has been discussed extensively elsewhere (Meredith et al 1983a, 1989b), the mechanisms by which a molecular ensemble can contribute to the susceptibility of an optical process include all degrees of freedom with characteristic frequencies near or below the frequencies of the process and their combinations. Thus, though optical difficulties are greater, the certainty of the correct partitioning of contributions of different mechanisms is greatest for simple harmonic generation, less for sum and difference mixing and most difficult for processes involving zero frequency fields or fields which have combinations which are zero valued (e.g. EFISH, electro-optic, Kerr processes).

2.1   Techniques for molecules in solutions

Having an interest in comparing the influences of variations of molecular structure on the electronic hyperpolarizabilities of molecular compounds we developed techniques which allow this to be accomplished in a reliable and efficient manner (Stevenson et al 1986, Meredith et al 1989a and b). As stated above, harmonic generation methods are associated with the most clearcut mechanisms of polarization. Further, being interested in the structural dependence of polarizability (and in colored compounds), rather than the influence on the distribution and character of low lying electronic excited states, we have chosen to use the longest conveniently accessible, intense light source with which harmonic generation could be performed: the hydrogen Raman-shifted Nd laser radiation near 1.91 microns. This requires some compromises since detectors for 954 nm second harmonic radiation pulses are not as efficient as for shorter wavelengths and since our preferred technique requires tight focussing, optical damage then setting a limit to the amount of light which can be used.

In order to measure many compounds in a timely manner, one is forced to use fluids, thus giving up the opportunity to probe the anisotropy of hyperpolarizability. Since vapor pressures are often low for larger molecules, one is also forced to use solutions, thus giving up a degree of accuracy due to the uncertainty in condensed-phase dielectric models. The two methods employed are optical third harmonic generation (THG) and dc-electric field induced second harmonic generation (EFISH).

THG is straightforward conceptually. Using linearly polarized light, a parallel polarized polarization oscillating at three times the fundamental is induced in isotropic media:

$$P^{3\omega} = \chi^{(3)}(-3\omega;\omega,\omega,\omega)_{1111} (E^{\omega})^3 \quad .$$

The susceptibility in an uncorrelated, mean-field aproximation is related, via number density and local field factors, to the scalar, or the 1111 component of the orientationally averaged projection, of the hyperpolarizability tensor, γ:

$$\chi^{(3)}(-3\omega;\omega,\omega,\omega)_{1111} \approx \Sigma_i \{ \rho^N (f^{\omega})^3 f^{3\omega} \gamma \}_i + \chi^{(3)}_{cascading} \quad .$$

The last term is a complication due to the possibility that highly nonlinear species might experience an indirect production of a 3ω polarization by sequential second-order polarization processes involving local electric fields (Meredith 1982, et al 1983b). Below that complication was diminished by use of low polarity sovents. The frequency dependent local field factors, f^{ω}, are often approximated by the Lorentz-Lorenz expression. The optical Onsager expression may also be used, which differs from the L-L function when the volume normalized linear polarizability of solvent and solute molecules differ. A high power of f^{ω} occurs in nonlinear susceptibilities and it can be easily shown that, in contrast to the conclusion of an earlier report (Singer et al 1981), there can thereby be a significant difference in the magnitudes of deduced hyperpolarizabilities. The Onsager model, disregarding cascading, was used for most of the hyperpolarizability results below.

Experiments where $\chi^{(3)}(-3\omega;\omega,\omega,\omega)_{1111}$ of a liquid can be precisely and, hopefully but not necessarily, accurately determined can be designed. The process involves consideration of field generation and coherent interference in a Type B process, with specific details of focussing in THG. This has been described extensively elsewhere (Meredith 1983). In summary, since every medium can generate a third-harmonic polarization, care must be taken to reduce the contributions from extraneous parts of the experimental apparatus such as air, lenses, mirrors, beam splitters, prisms and filters. Several methods have been used. We currently use a cell with large input window and sample compartment where tight focussing at their interface produces a "constant" signal which (limiting application to real and nearly real susceptibilities) is to within some Fresnel transmission factors proportional to the differences of $\{\chi^{(3)}(-3\omega;\omega,\omega,\omega)_{1111}/[\chi^{(1)}(-3\omega;3\omega)_{11}-\chi^{(1)}(-\omega;\omega)_{11}]\}$ in the liquid and the window. There are several advantages to this. First, the signal is very insensitive to temperature, a potential problem due to weak absorption at 1.91 microns in most liquids. Second, the signal is the resultant from the "subtractive" interference between the window and liquid responses. Thus, the precision of the technique is improved over that of an absolute measurement. This is significant since pulse-to-pulse laser energy and mode jitter cause large THG fluctuations. A variation of concentration method is used to extract hyperpolarizabilities at the lowest possible concentrations. Therefore, since solvents have "normal" nonlinearity and since most materials produce a surprisingly constant level of non-phasematched THG, the differential method consistently boosts precision. The dispersion denominators required to extract $\chi^{(3)}$'s are easily measured by the simple wedge method. An advantage to this approach over the use of a single cell with smaller liquid chamber in the shape of a wedge with a long output window, is that the fit of the resultant sinusoidal nonlinear fringe pattern entails four or more parameters (amplitude, period, phase, offset, attentuation constants). Inexact data results in a play-off between parameters to achieve the best fit. By separately determining amplitudes and dispersion, this problem is avoided.

There remains the issue of calibrating the magnitude of the detected response. This is done by referencing. First, the pulse-to-pulse jitter and experimental apparatus drift can be normalized by taking ratios to a signal generated in an equivalent, static reference arm. Second, this "stabilized" signal can be referenced to the response of a secondary standard in the sample arm. Two things argue for using as secondary standard another liquid placed in the other half of a side-by-side sample cell. Using a standard of different physical format introduces changes in the optical apparatus. There are many sensitive factors in the efficiency of THG generation and detection, such as the position relative to the focus, beam profile changes due to nonlinear refraction, the spatial or intensity dependent variation of photocathode efficiency, etc. It is obvious that highest precision results from use of "identical" configurations and nearly equivalent media Second, the apparatus will drift in ways that the reference arm does not correct. By rapidly switching the cell between sample and secondary standard, a greater precision can be achieved than when the comparison is made once or twice at the beginning and/or end of the sample measurement. Calibration of the secondary standard is the determination of it's response relative to glass, which has been extracted from earlier work employing a different methodology (Meredith et al 1983a), using a statistical criterion to improve the confidence over that of a single liquid (Stevenson et al 1986). The glass nonlinearity was referenced to crystalline quartz (Buchalter et al 1982) and the quartz nonlinearity was determined from details of the nonlinear fringe pattern which results from interference of direct THG with two-step, cascaded SHG/THG in the same crystal (Meredith 1981b). Though widely quoted, the author of this original work has questioned the accuracy due to either multimodal effects in cascading or the accuracy of the second-order d coefficients whose square was used to calibrate $\chi^{(3)}(-3\omega;\omega,\omega,\omega)_{xxxx}$. Nevertheless, since the calibration route of our techniques only utilize this as the primary standard, any improvements will simply scale all THG results proportionally.

EFISH in fluids has been used extensively for extracting a component of the β tensor, but it is less clear in liquids that this is accurately accomplished. Nevertheless, it is a valuable tool. As in THG, for EFISH using linearly polarized light, a parallel polarized polarization oscillating at twice the fundamental is induced in isotropic media:

$$P^{2\omega} = \chi^{(3)}(-2\omega;0,\omega,\omega)_{1111} E_0(E^\omega)^2 \quad .$$

As for THG the susceptibility is related to the scalar γ, plus a term due to the perturbation of the isotropic orientational distribution function which projections second-order hyperpolarizability to the macroscopic level:

$$\chi^{(3)}(-2\omega;0,\omega,\omega)_{1111} \approx \Sigma_i \{ \rho^N (f^\omega)^2 f^{2\omega} [f^0\gamma + f^{0'}\mu\beta/5KT] \}_i + \chi^{(3)}_{cascading} \quad .$$

Remarks above concerning cascading and f^ω are also appropriate here. The $\mu\beta$ factor is by now a well known abreviation (Singer et al 1981). The zero-frequency local field factors are often taken in the Onsager model, where the influence of dipole moments on determining reorientation suggests the two terms may differ. Kirkwood-correlation could also be included, though the use of dipoles measured in the same solvent is thought to mitigate the error. The low and high frequency Onsager models were used for results listed below. The γ, or "deformation term", of EFISH differs from the γ of THG both in the dispersion of the electronic contribution and in the larger hybrid vibrational-electronic contribution. Often the influence of γ is ignored. We attempt to include its effect by scaling THG results under the assumption that $\{\gamma_{deform}/\gamma_{THG}\}$ can be treated as a constant which can be determined in molecules not possessing permanent dipoles. This is not correct, but probably better than ignoring the effect.

Many features of EFISH optics parallel THG with many of the same problems (Meredith 1981a, 1983). It is advantageous that EFISH only arises in regions of large dc fields and that the sense and magnitude of $\chi^{(2)}_{eff} = \chi^{(3)}(-2\omega;0,\omega,\omega)\cdot E_0$ is controlled by the polarity and magnitude of E_0. This can be used, obviously, for periodic poling, but also allows a route to intracell calibration. To gain the advantages listed above, electrodes were applied to the windows of the THG cell allowing concurrent detection of EFISH. Since in the modified cell the E_0 are identical at the interface, and an adiabatic SHG generation occurs, the detected response is proportional to the differences of $\{\chi^{(3)}(-2\omega;0,\omega,\omega)_{1111}/[\chi^{(1)}(-2\omega;2\omega)_{11}-\chi^{(1)}(-\omega;\omega)_{11}]\}$ in the liquid and the window. Precision in dispersion, or coherence length, is increased using crystalline quartz windows on the simple wedged cell, which is used simultaneously for THG. Thus, instead of measuring the dispersion of EFISH directly, the period of variably retarded interference of two SHG signals is seen. The dispersion of the liquid provides the same thickness dependent retardation as would occur in EFISH (diregarding the negligible difference due to differences in quadratic electro-optic effect at ω and 2ω). The advantage is that the SHG signal is orders of magnitude stronger. The calibration of the secondary standard is accomplished relative to crystalline quartz. For this the front window of the dual chamber cell is replaced with an oriented crystal window to which electrodes are attached. Considerations of the SHG under variable fields with and without liquids allows calibration (Stevenson 1988).

Other quantities (n, ε, ρ) are measured, with commensurate accuracy and precision, so that precise susceptibilities are determined allowing reliable hyperpolarizability deduction.

2.2 Techniques for thin films

For molecular thin films and thicker polar-oriented polymeric films SHG, ellipsometry, waveguiding and electro-optic characterization provide assessment of the nonlinearity and evolution of these often nonequilibrium assemblies (Meredith et al 1989b).

3.   MOLECULAR HYPERPOLARIZABILITY

The THG and EFISH techniques described above have been used to characterize a wide range of materials. Some of these studies have been described elsewhere (Meredith et al 1989a, Cheng et al 1989). A theme which was identified, e.g. in the study of naphthalene derivatives, was the reduced enhancement of β for the classical resonance capable isomers over other isomers. This raised again the question of the origins of enhancement in larger molecules. The two level model has been widely quoted (Meredith 1988). The underlying premise is that the regions of electron donating and electron accepting strengths are displaced and there is one intense, low lying optical transition which moves charge substantially between them. The implication is that the remainder of the molecule's excitations do not collectively compete. In larger molecules, such as the naphthalenes, we suggest this is untrue. Clearly the larger size leads to a denser set of excited states with the consequence that the polarizability of the system is the result of an easier biasing of the whole system. If one expects to see the two-level model at work, it should be in the smaller molecules where the polarization processes are more difficult due to the quantum mechanical consequences of smaller size, that is, higher kinetic energy costs for wavefunction distortions. In this regard, the results of some empirical correlations illustrated below are interesting.

In the course of creating an empirical database, a few things were realized that warrant mention. First, there is a level of uncertainty brought in by the models of local fields and the necessity to include deformation terms. The attempt to include the EFISH deformation by the scaling of THG, as described above, results in values of the $\mu\beta$ orientation term which are often too small to be reliable. This will obviously occur since molecules span the range from centrosymmetric with vanishing β to highly asymmetric with very large β. What is surprising, in view of results reported in earlier literature, is that significant uncertainty occurs even for many monosubstituted benzenes. When approximated in this manner the deformation term is never negligible. It rarely falls significantly below ten percent of the orientational term, even for large β molecules (Cheng et al 1989).

Returning to polarization mechanisms, if low-lying charge-transfer excitations are responsible for the large nonlinearities in para donor-acceptor substituted benzenes, there should be a trade-off between nonlinearity and transparency. To explore this question, we have characterized a wide range of them, including highly colored dicyanovinyl and tricyanovinyl derivatives and transparent cyano and aldehyde derivatives. Measurement results, showing good correlation betwen β and λ_{max} of the lowest optical transition, in Fig. 1 confirm that such a trade-off exists, at least in benzenes. Since measurements are made with 1.91 μm radiation, the increase is not the result of dispersive enhancement above the hypothetical zero-frequency electronic β due to finite ω and 2ω. The question then is whether the widely quoted two-level model can account for this correlation. For long-axis polarizations and $\omega \sim 0$, $\beta_{TLM} \propto (\lambda_{max})^3 f \Delta\mu$, the product of cubed wavelength with the oscillator strength and difference of Franck-Condon-state dipole moments. There is no known systematic dependence of f and $\Delta\mu$ on λ_{max}, therefore critical assessment awaits further spectroscopic and solvatochromic measurements. One might expect f and $\Delta\mu$ within this class of weakly polarized molecules to be correlated with the combined donor-acceptor

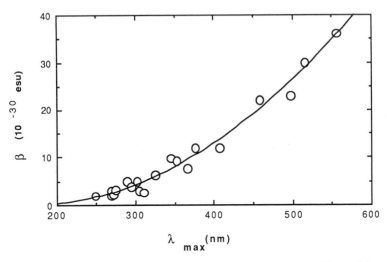

Fig. 1. Correlation of the low frequency hyperpolarizability β with λ_{max} of the
lowest optical absorption feature for p-disubstituted benzenes.

strength, which factor might also increase λ_{max}. Allowing for an unidentified positive correlation which is here approximated as f $\Delta\mu \propto (\lambda_{max})^m$ with m≥-1 (since f $\propto \mu_{ge}^2/\lambda$) the two level model predicts $\beta_{TLM} \propto (\lambda_{max})^n$, n=3+m. However, oscillator strengths of these bands have been measured and exhibit no correlation with wavelength, typically varying between 0.3 and 0.7, thus requiring m≥0, n≥3. This prediction is in agreement with the data as represented in Fig. 2 where in a log-log plot, within the uncertainty of some dispersion enhancement of larger λ_{max} data, a near quartic variation of β with λ_{max} occurs (best line slope = 3.8). Therefore this series is consistent with the two-state model. Descriptions of variations in other series and tests of other mechanisms of nonlinearity are forthcoming (Cheng et al 1989).

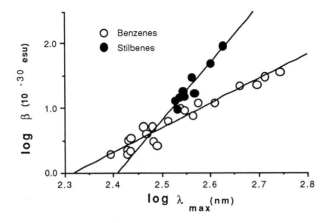

Fig. 2. Logarithmic plot of β vs λ_{max} for p-disubstituted benzenes (open circles)
and 4-4'-disubstituted stilbenes (filled circles). Best fit lines are included.

Table 1. Hyperpolarizabilities of 4-methoxy-4'-nitroarenes.

Structures β $(10^{-30}esu$)

Structure	β $(10^{-30}esu)$
H_3CO—⟨benzene⟩—NO_2	5.1
H_3CO—⟨benzene⟩—⟨benzene⟩—NO_2	11
H_3CO—⟨benzene⟩≡⟨benzene⟩—NO_2	16
H_3CO—⟨benzene⟩═⟨benzene⟩—NO_2	29
H_3CO—⟨benzene⟩—⟨benzene⟩—⟨benzene⟩—NO_2	10
H_3CO—⟨benzene⟩—⟨benzene⟩═⟨benzene⟩—NO_2	18
H_3CO—⟨benzene⟩═⟨benzene⟩═⟨benzene⟩—NO_2	35

Effects of conjugation length and planarity on nonlinear polarizabilities are illustrated in 4,4' disubstituted biphenyls, dipenyl acetylenes, trans-stilbenes, and other phenyl-phenylvinylenes. Results for only the 4-methoxy-4'-nitro derivatives are given in Table 1. As a result of extended conjugation, biphenyl derivatives give higher β values than their benzene analogues. The extent of enhancement however depends on the strength of donor and acceptor groups. Since the phenyl rings are free to rotate, breaking up the conjugation pathway for electron polarization, the biphenyl structure cannot be highly effective for optical nonlinearity. However, the degree of planarity is also expected to depend on the electron donating and accepting strength of the substituents. With strong resonance donors such as amine and dimethylamine, substantial double bond character may be imposed on the σ linkage of biphenyl, leading to a more planar structure and higher molecular nonlinearity.

Its extended, planar structure makes trans-stilbene a very effective conjugated backbone. In general between analogous benzene and stilbene structures, typical enhancement of 6 to 8 times is realized for the molecular hyperpolarizability (Cheng et al 1989). Such is the case even for the donor or acceptor monosubstituted derivatives. Only modest increases, typically between 0.5 to 0.8 Debye, are seen for the ground state dipole moments. With the substantial increase in conjugation length, this indicates a decrease in the fractional charge formally transfered among groups in the ground state. In addition, the charge-transfer bands of trans-stilbene derivatives are typically 50 to 60 nm red shifted in comparison to that of the benzenes. A trade-off between nonlinearity and transparency is also observed. For a given λ_{max}, however, substantially higher nonlinearity is offered by the stilbene derivatives. Fig. 2 shows a much stronger dependence of β on λ_{max}, in this case probably a charge-transfer band, than that of the benzene system.

It is often suggested that it is generally possible to obtain higher nonlinearity with longer conjugation. Examination of longer arene derivatives, however, shows otherwise. Substantial decreases of nonlinearity per unit length are found in longer derivatives (Table 1). In the case of terphenyl, nonplanarity must account for some of the decrease. The modest β of the expectedly planar phenylvinylene stilbene derivative is somewhat puzzling. Bond alternation, as a result of the intrinsic asymmetry of the end groups, may account for part of the decrease. This however cannot be the principle reason since no saturation effect is observed for the enhancement of β with conjugation length for the polyethylenic diphenyl derivatives (Huijts et al 1989) and push-pull polyenes (Barzoukas et al 1989) of comparable lengths. An important property of the conjugation structures in Table 1 is their formal aromaticity according to the 4n+2 rule. The consequence of effective charge transfer interaction includes the conjugation of the donor electron density, thus resulting in a significant deviation from the aromatic configuration. This apparent trade-off between charge transfer and aromaticity may help rationalize the rapid saturation of nonlinearity vs length for arene derivatives.

4. CRYSTALS

Having conducted a powder SHG survey of many organic compounds, interesting cases and classes have been discovered. The search was partly guided by the need for increased transparency in the blue, near ultraviolet and near infrared. Some of these materials will shortly be disclosed and discussed (Donald et al 1989). It is interesting to investigate the cases of the few molecules which align significantly, comparing them to similar structures which fail. For example, a study of 4-halobenzonitriles was instigated by our SHG probing. This has lead to the recognition not only of the metastability of the literature crystal structure of chlorbenzonitrile, but more significantly the importance of halo-cyano interactions in determining crystal structures (Desiraju et al 1989).

Recently we have also used SHG as a probe of molecular orientation in single crystals grown with significant occlusion of guest species. The conjecture was that crystal symmetry could be reduced, and thereby SHG activity induced, since the guests would be incorporated through growing faces. Since faces have lower symmetry than interiors, the occluded crystal would thus reflect the reduced symmetry. This was seen in the elimination of symmetry-required zeros of the d tensors which also reflected other aspects of facet growth (Weissbuch et al 1989).

5. MOLECULAR THIN FILMS

The use of optical techniques to probe the behavior of molecular thin films is quite powerful. High sensitivity ellipsometry of the additional retardation imparted at the air-water interface by Langmuir films enables real time probing of molecular behavior on troughs with submonolayer sensitivity (Meredith et al 1986). An example of such behavior is illustrated in Fig. 3 where the additional retardation from fully compressed monolayers of fatty alcohols and acids is depicted. The linear dependencies on alkyl chain length is not surprising, but the zero retardation intercepts in the vicinity of n ~ 8 is. One should remember that shorter compounds do not form monolayers due to their solubility in water. It is tempting, but premature, to speculate on the nature of interpenetration of water and surfactant layers.

We have earlier described the use of a strategy to force Z-type deposition due to alteration of the hydrophobicity of the alkyl tails of surfactants by inclusion of amide groups (Popovitz-Biro et al 1988). SHG was used as direct proof of the formation and stability of such multilayers for surfactant molecules carrying terminal p-nitroanilino nonlinear "markers". For use as nonlinear media, misorientation of the "marker" from the film normal must be reduced.   In this regard, the dependence of the orientation on spacings within the hydrophobic tail of these compounds has been investigated (Hsiung et al 1989).   An example of the determination of this orientation and the variation between surfactants differing by one methylene group is presented in Fig. 4.   The technique relies on the variation of p-polarized SHG intensity on the polarization angle of the incident fundamental.   Fundamentally the ratio of two components of the surface   is measured, but in a fashion allowing greater finesse over sequential pure   polarization   measurements.

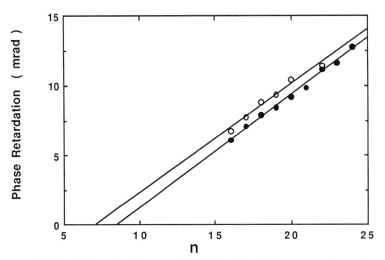

Fig. 3. Additional surface retardation induced by fully compressed monolayers of fatty acids (solid) and alcohols (circles) at the air-water interface. There are n carbon atoms in a molecule.

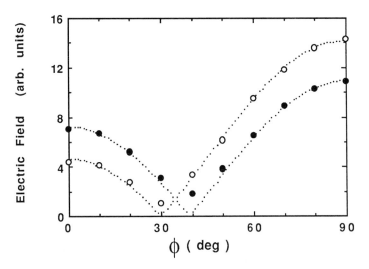

Fig. 4. Amplitude of p-polarized second harmonic as a function of excitation polarization for compressed monolayers at the air-water interface. Dotted curves are theoretical fits. Dots (circles) are compound 2 (1).

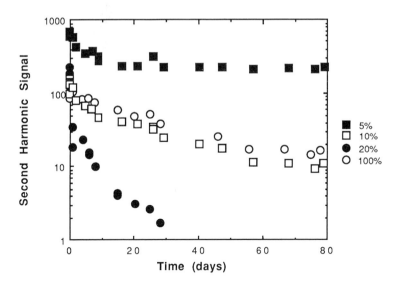

Fig. 5. Relaxation of SHG signal from thin films of corona-poled pendant-group polymers. Fraction of pendant positions occupied by disperse red dye is varied.

Polymeric materials containing species with significant β have been of interest for several years. The original premise (Meredith et al 1983) was that some nonlinearity which might be achieved in well aligned media could be traded for flexibility of polymers in which alignment of a lesser degree would be achieved by the action of poling fields. Pendant-group polymers of many types have been prepared for this purpose. Details concerning the stability of orientation in such polymers are not well known. An example of the ability to observe the relaxation over significant times with the use of an SHG apparatus is illustrated in Fig. 5. We do not comment here on the factors affecting the various rates of relaxation as functions of time or fractional coverage of attachment sites.

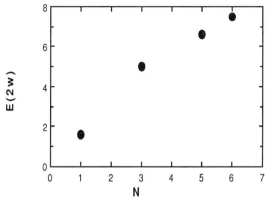

Fig. 6. Amplitude of second harmonic from multilayer LB films as function of layer number N. The polymer is similar to those of Fig. 5.

The alignment of pendant groups in these polymers can also be induced at the air-water interface of a Langmuir trough. Fig. 6 shows that it can be maintained to great extent when stacking layers via the dipping transfer method. Maintenance of identical alignment in all layers would be ideally indicated by a linear dependence of the SHG amplitude on number of deposited layers. Here the deviation from linearity is not great.

CONCLUSION

Methods which are used in our laboratories for study of nonlinearity and molecular assembly and orientational evolution have been reviewed. Examples of their use have also been reviewed or presented. In our opinion there remain many aspects of molecular nonlinear hyperpolarizability to be uncovered. Also, realizing that molecular information is insufficient for technological advancement we expect that convenient and reliable probes of molecular organization will provide vital tools in the invention and development of reliable thin film and polymer media for nonlinear optics.

ACKNOWLEDGEMENTS

The authors thank their numerous collaborators, particularly members of The Weizmann Institute of Science and of Philips Research Laboratories.

REFERENCES

Barzoukas M, Blanchard-Desce M, Josse D, Lehn J-M , and Zyss J, 1989 *Chem. Phys.* **133** 323
Buchalter B and Meredith G R 1982 *Appl. Opt.* **21** 3221
Cheng L-T, Tam W, Meredith G R, Rikken G L J A and Meijer E W, 1989 *SPIE Proc.* **1147** to appear
Desiraju G R and Harlow R L 1989 *J. Am. Chem. Soc.* to appear
Donald D S, Cheng L-T, Desiraju G, Meredith G R and Zumsteg F C 1989 to appear
Hsiung H, Meredith G R, Vanherzeele H, Popovitz-Biro R and Lahav M 1989 *Chem Phys Lett* submitted
Huijts R A and Hesselink G L J 1989 *Chem. Phys. Lett.* **156** 209
Meredith G R 1981a *Opt. Commun.* **39** 89
Meredith G R 1981b *Phys. Rev. B* **24** 5522
Meredith G R 1982a *Chem. Phys. Lett.* **92** 165
Meredith G R 1982b *J. Chem. Phys.* **77** 58638
Meredith G R 1983 *Nonlinear Optical Properties of Organic and Polymeric Materials* ed D J Williams (Washington D C: Am. Chem. Soc.) pp 27-56
Meredith G R, Vandusen J G and Williams D J 1983 *Nonlinear Optical Properties of Organic and Polymeric Materials* ed D J Williams (Washington D C: Am. Chem. Soc.) pp 110-133
Meredith G R, Buchalter B and Hanzlik C 1983a *J. Chem. Phys.* **78** 1533; and 1543
Meredith G R and Buchalter B 1983b *J. Chem. Phys.* **78** 1938
Meredith G R 1987 *SPIE Proc.* **824** 126-133
Meredith G R 1988 *Mat. Res. Soc. Bull.* (**August**) 24-29
Meredith G R and Stevenson S H 1989a *Nonlinear Optical Effects in Organic Polymers* ed J Messier et al (Kluwer) pp 105-122
Meredith G R, Hsiung H, Stevenson S H, Vanherzeele H and Zumsteg F C 1989b *Organic Materials for Nonlinear Optics* ed Hahn R A and Bloor D (London: The Roy. Soc. of Chem.) pp 97-111
Popovitz-Biro R, Hill K, Landau E M, Lahav M, Leiserowitz L, Sagiv J, Hsiung H, Meredith G R and Vanherzeele H 1988 *J. Am. Chem. Soc.* **110** 2672
Singer S H and Garito A F 1981 *J. Chem. Phys.* **75** 1533
Stevenson S H and Meredith G R 1986 *SPIE Proc.* **682** 147-152
Stevenson S H 1988 unpublished
Weissbuch I, Lahav M, Leiserowitz L, Meredith G R and Vanherzeele H 1989 *Chem. Mater.* **1** 114

Inst. Phys. Conf. Ser. No 103: Section 2.2
Paper presented at Int. Conf. Materials for Non-linear and Electro-optics, Cambridge, 1989

A simple method for the determination of $\chi^{(3)}$-coefficients based on the DC Kerr effect

H M M Klein Koerkamp, T H Hoekstra, G J M Krijnen, A Driessen, P V Lambeck and Th J A Popma.

Faculty of Electrical Engineering, Transducers and Materials Science, University of Twente, P.O. BOX 217, 7500 AE Enschede, The Netherlands.

ABSTRACT: A method will be presented for the determination of the third-order nonlinear susceptibility $\chi^{(3)}(-\omega;0,0,\omega)$ using a simple multilayer structure provided with thin metal electrodes. The low frequency field-induced refractive-index changes are determined by a dynamic measuring method based on the refractive index dependence of the phase matching condition for coupling light into waveguiding structures. Experiments showed that $\chi^{(3)}(-\omega;0,0,\omega)$ values in the order of $10^{-21} m^2/V^2$ could be determined with a probe beam of 5mW and a wavelength of 632.8 nm (HeNe).

1. INTRODUCTION

All optical signal processing requires materials which posses high nonlinear $\chi^{(3)}$-coefficients and meet the requirements for use in integrated optic devices like transparency and the possibility of thin film deposition. In the development of these materials a simple technique is needed to determine the $\chi^{(3)}$-coefficients. Methods like nonlinear grating (Chen *et al.* 1986) and prism coupling (Patela *et al.* 1986), interferometry (Lattes *et al.* 1983), degenerate four-wave mixing (Rao *et al.* 1986) and third harmonic generation (Hermann 1973) require high optical powers and often laborious sample preparation. When determining the nonlinear susceptibility at optical frequencies, problems arise because the field distribution is not known well enough. If however the nonlinearity is merely electronic in origin, the low frequency $\chi^{(3)}$ is comparable with the nonresonant optical $\chi^{(3)}$ and can be determined from the induced refractive index change by an applied dc or low frequency electric field. This is known as the DC Kerr effect.

2. DC KERR EFFECT

The nonlinear polarization can formally be written as:

$$P(\omega) = \varepsilon_0\chi^{(1)}(\omega).E(\omega) + \varepsilon_0\sum_{\omega_a,\omega_b} \chi^{(2)}(-\omega;\omega_a,\omega_b):E(\omega_a)E(\omega_b) +$$

$$\varepsilon_0\sum_{\omega_a,\omega_b,\omega_c} \chi^{(3)}(-\omega;\omega_a,\omega_b,\omega_c):E(\omega_a)E(\omega_b)E(\omega_c) \qquad (1)$$

where the summation over ω_a, ω_b and ω_c contains the positive and negative frequencies of all components in the total electric field E. In the following we shall restrict ourselves to third-order nonlinearities. The nonlinear part of the polarization is then given by (Maker and Terhune 1965):

$$P_i^{NL}(\omega_d) = \varepsilon_0 . \sum_{\omega_a, \omega_b, \omega_c} D . \chi_{ijkl}^{(3)}(-\omega_d; \omega_c, \omega_b, \omega_a) . E_j(\omega_c) . E_k(\omega_b) . E_l(\omega_a) \qquad (2)$$

$$\text{with} \quad \omega_a + \omega_b + \omega_c - \omega_d = 0$$

where the Einstein summation convention is used and $D = 6/n!$ is the degeneracy factor with n the number of identical frequencies. The degeneracy factor D accounts for the permutation of the electric field amplitudes. A field may enter the product EEE as the conjugate of the other. Their frequencies are nonidentical because the conjugated field can be expressed as a field with negative frequency:

$$E^*(\omega_a) = E(-\omega_a) \qquad (3)$$

For isotropic media the $\chi^{(3)}$-tensor contains 81 elements of which 21 are nonzero and due to Kleinman symmetry the number of independent elements is reduced to only one (Hopf and Stegeman 1986):

$$\chi_{ijkl}^{(3)} = \frac{1}{3} \chi_{1111}^{(3)} . (\delta_{ij}\delta_{kl} + \delta_{ik}\delta_{jl} + \delta_{il}\delta_{jk}) \qquad (4)$$

In the case of the DC Kerr effect two of the three fields are externally applied dc electric fields and equation (2) becomes:

$$\begin{aligned}
P_i^{NL}(\omega) = \quad & 3 . \varepsilon_0 . \chi_{ijkl}^{(3)}(-\omega; 0, 0, \omega) . E_j(0) . E_k(0) . E_l(\omega) \qquad (5) \\
+ \; & 3 . \varepsilon_0 . \chi_{ijkl}^{(3)}(-\omega; 0, 0, \omega) . E_j^*(0) . E_k^*(0) . E_l(\omega) \\
+ \; & 6 . \varepsilon_0 . \chi_{ijkl}^{(3)}(-\omega; 0, 0, \omega) . E_j^*(0) . E_k(0) . E_l(\omega) \\
= \; & 12 . \varepsilon_0 . \chi_{ijkl}^{(3)}(-\omega; 0, 0, \omega) . E_j(0) . E_k(0) . E_l(\omega)
\end{aligned}$$

where $E_j^*(0) = E_j(0)$ is used. The nonlinear part of the polarization can be seen as a perturbation of the linear polarization:

$$\begin{aligned}
P_i(\omega) &= (n_i^2 - 1) \, E_1(\omega) \qquad (6) \\
&\approx (n_{i0}^2 - 1) \, E_1(\omega) + 2n_{10}\Delta n_i \, E_1(\omega) \\
&= P_i^L(\omega) + P_i^{NL}(\omega)
\end{aligned}$$

$$\text{with} \quad n_i = n_{i0} + \Delta n_i$$

With equations (5) and (6) the change in the refractive index which can be associated with the nonlinear polarization can be written as:

$$\Delta n_i = \frac{6}{n_{10}} \chi_{ijki}^{(3)}(-\omega; 0, 0, \omega) . E_j(0) . E_k(0) \qquad (7)$$

If the dc electric field is parallel to the x-axis the refractive index changes along the various axes are given by:

$$\Delta n_x = \frac{6}{n_{x0}} \chi_{1111}^{(3)}(-\omega; 0, 0, \omega) . E_{dc}^2 \qquad (8a)$$

$$\Delta n_y = \frac{2}{n_{y0}} \chi_{1111}^{(3)}(-\omega; 0, 0, \omega) . E_{dc}^2 \qquad (8b)$$

$$\Delta n_z = \frac{2}{n_{z0}} \chi_{1111}^{(3)}(-\omega; 0, 0, \omega) . E_{dc}^2 \qquad (8c)$$

Formulas (8a), (8b) and (8c) give a direct relation between the refractive index change induced by a dc electric field and the $\chi^{(3)}_{1111}$-coefficient.

3. MEASUREMENT METHOD

The reflectance R in an Attenuating Total Reflection (ATR) set-up depends on the efficiency of coupling light into lossy waveguiding structures, and is controlled by the phase mismatch δ:

$$\delta = k_{z,prism}(\theta) - k_{z,mode} \qquad (9)$$

with k_z the tangential component of the propagation vector. The reflectance R is minimal (R_{min}) when phase matching occurs ($\delta=0$), which can be realized by adjusting the external angle θ. A variation in the refractive index of one of the layers causes a variation in $k_{z,mode}$ and results in a change of the reflectance R. In order to measure index variations of about 10^{-7} we developed a dynamic measuring method. By applying a low frequency electric field $E\sin(\omega_m t)$ the refractive index of the nonlinear material can be modulated. Because the refractive index variations (eq.8) are proportional with E^2, the modulation frequency is $2\omega_m$. The modulated refractive index causes a modulation in the reflectance R. This differential reflectance R_{diff} can be monitored with a lock-in technique:

$$R^{TE}_{diff}(\theta) = \left[\frac{\partial R(\theta)}{\partial n_y}\right]_\theta \cdot \Delta n_y \qquad (10a)$$

$$= \left[\frac{\partial R(\theta)}{\partial n_y}\right]_\theta \cdot \frac{1}{n_{y0}} \chi^{(3)}_{1111}(-\omega;0,0,\omega).E^2(1+\sin(2\omega_m t))$$

$$R^{TM}_{diff}(\theta) = \left[\frac{\partial R(\theta)}{\partial n_x}\right]_\theta \cdot \Delta n_x + \left[\frac{\partial R(\theta)}{\partial n_z}\right]_\theta \cdot \Delta n_z \qquad (10b)$$

$$= \left[\left[\frac{\partial R(\theta)}{\partial n_x}\right]_\theta \cdot \frac{3}{n_{x0}} \chi^{(3)}_{1111}(-\omega;0,0,\omega) + \right.$$

$$\left. \left[\frac{\partial R(\theta)}{\partial n_z}\right]_\theta \cdot \frac{1}{n_{z0}} \chi^{(3)}_{1111}(-\omega;0,0,\omega)\right].E^2(1+\sin(2\omega_m t))$$

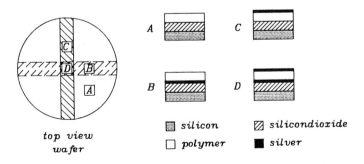

top view
wafer

Figure 1. Wafer geometry. The structures A, B, C and D differ in the number of silver layers but are otherwise identical. Structure D is used for the determination of the $\chi^{(3)}$-coefficients.

LAYER	ñ	d [μm]
silver (top)	0.0652+j4.47	0.052
polymer	1.619	1.165
silver (bottom)	0.0652+j4.15	0.050
silicondioxide	1.454	~ 1

Table 1. Complex refractive index ñ and thickness d of the various layers in the structure.

The factor $(\partial R/\partial n_i)$ in the formulas (10a) and (10b) can be written as:

$$\left[\frac{\partial R(\theta)}{\partial n_i}\right]_\theta = \left[\frac{\partial R(\theta)}{\partial \theta}\right]_{n_i} \cdot \left[\frac{\partial \theta}{\partial n_i}\right]_R \tag{11}$$

For small refractive index changes the shape of the individual ATR-dips is conserved:

$$\left[\frac{\partial \theta}{\partial n_i}\right]_R = \left[\frac{\partial \theta}{\partial n_i}\right]_{R_{min}} = \left[\frac{\partial \theta_{min}}{\partial n_i}\right] \tag{12}$$

where $(\partial \theta_{min}/\partial n_i)$ is mode-dependent and accounts for the change in the angle of minimal reflectance due to a refractive index change. This factor can be determined from simulations. The shape of the dips is incorporated in the factor $(\partial R/\partial \theta)$, which can be taken from measurements.

4. RESULTS

The geometry of the layer structure we used is given in Figure 1. The silicondioxide layer is thermally grown on a 2-inch silicon wafer. The silver layers are deposited by vacuum evaporation and the nonlinear polymer layer by spin coating. The polymer is an amorphous side-chain polymer incorporating methoxy-nitro-stilbene (MONS) groups and is developed by Akzo Corporate Research Labs for high $\chi^{(2)}(-\omega;0,\omega)$ coefficients (Horsthuis and Krijnen 1989). The nonlinear properties of the polymer are due to the conjugated π-electron systems and are, apart from thermal effects, electronic in origin. Patterning of the silver layers, which are to be used as electrodes, resulted in the realization of four

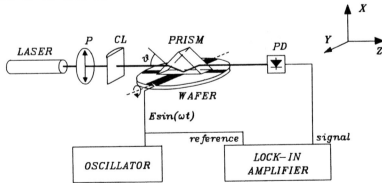

Figure 2. Experimental set-up. The refractive indices of the used rutile prism were: $n_0=2.584$(TM-modes), $n_e=2.872$(TE-modes) and the prism angle was $\alpha_p=44.47^0$.
P: polarizer; CL: cylindrical lens; PD: photodiode.

different structures. The refractive index and thickness of the polymer
layer are determined by prism coupling (structure A). The other layers are
characterized by fitting the measured angles of minimal reflectance θ_{min}
(structures B, C and D) with the ones calculated with a simulation program
(table 1). The difference in the imaginary part of the refractive indices
of the top and bottom silver layer is caused by a chemical reaction of the
top layer with the surrounding atmosphere. The experimental set-up is
given in figure 2. The laser was a 5mW HeNe-laser (λ=632.8 nm), the
frequency of the applied electric field 5 kHz and the lock-in amplifier
operated in the 2f-mode. The detection system showed a flat frequency
response up to 20 kHz. In figures 3 and 4 the measured ATR-curves and
differential reflectance $R_{diff}(\theta)$ for the TE and TM modes belonging to
structure D of figure 1 are given. Except for dip TM-3 all the other dips
were used for an independent determination of the $\chi_{1111}^{(3)}$-coefficient
(table 2). In calculating the $\chi_{1111}^{(3)}$-coefficient (eq.10) the factor
$(R_{diff}(\theta)/(\partial R(\theta)/\partial n_1))$ is necessary and therefore the system transfer is
cancelled out. Hence, the reflectance R and differential reflectance R_{diff}
are expressed in volts, making calibration unnecessary. Due to the
resolution of 0.01 deg. in the external angle θ the calculated values for
$(\partial R/\partial \theta)_{max}$ are too small for sharp dips e.g. TM-2, TE-1 and TE-2, leading
to values for $\chi_{1111}^{(3)}$ which are too high.

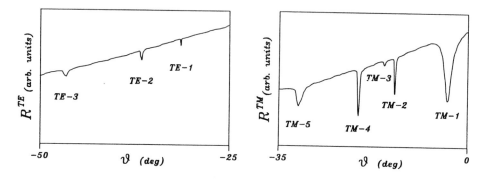

Figure 3. Measured ATR-curves for TE and TM modes.

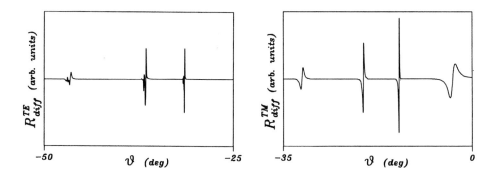

Figure 4. Measured differential reflectance for TE and TM modes.

No.	θ_{min} (deg)	$\left[\dfrac{\partial R}{\partial \theta}\right]_{max}$ (V/deg)	E_p (10^7V/m)	R^{max}_{diff} (V)	$\dfrac{\partial \theta_{min}}{\partial n_x}$ (deg)	$\dfrac{\partial \theta_{min}}{\partial n_y}$ (deg)	$\dfrac{\partial \theta_{min}}{\partial n_z}$ (deg)	$\chi^{(3)}_{1111}$ (m^2/V^2)
TM-1	-5.11	1.1	1.14	$6.9\ 10^{-5}$	86.1	–	11.9	$2.9\ 10^{-21}$
TM-2	-17.53	4.5	1.14	$2.5\ 10^{-4}$	69.7	–	4.6	$3.2\ 10^{-21}$
TM-3	-20.09	–	–	–	10.6	–	1.2	–
TM-4	-26.26	2.9	1.14	$1.3\ 10^{-4}$	66.8	–	18.3	$2.5\ 10^{-21}$
TM-5	-40.49	0.8	1.14	$4.3\ 10^{-5}$	66.9	–	44.8	$2.7\ 10^{-21}$
TE-1	-31.36	2.3	1.14	$5.4\ 10^{-5}$	–	80.0	–	$3.6\ 10^{-21}$
TE-2	-36.64	2.1	1.14	$4.8\ 10^{-5}$	–	86.9	–	$3.3\ 10^{-21}$
TE-3	-46.60	1.2	1.14	$2.6\ 10^{-5}$	–	105.3	–	$2.6\ 10^{-21}$

Table 2. Measured $\chi^{(3)}_{1111}$-coefficients in Akzo AFR18 polymer.

5. CONCLUSIONS

The presented method for the determination of $\chi^{(3)}$-coefficients is simple and needs no laborious sample preparation. Patterning of the silver layers realized four structures which made characterization of the various layers possible. The amplitude of the low frequency electric field in the nonlinear polymer layer was easy to determine due to the simple electrode geometry. Excitation of various modes resulted in a number of independent calculations of the $\chi^{(3)}_{1111}$-coefficient. The variations in the calculated values are mostly due to insufficient resolution in the external angle θ of 0.01 deg. as compared with the sharpness of especially the TM-2, TE-1 and TE-2 dips, which introduced larger errors in the determination of $(\partial R/\partial \theta)_{max}$ for these dips. Due to the detection technique the sign of $\chi^{(3)}_{1111}$ is lost. Because there was no electrical dissipation thermal effect could be excluded and the $\chi^{(3)}$ is purely electronic. The measured value for $\chi^{(3)}_{1111}(-\omega;\omega,0,0)$ was $3.0\pm0.6*10^{-21}$ m^2/V^2.

6. REFERENCES

Chen Y J, Carter G M, Sonek G J and Ballantyne J M 1986 Appl. Phys. Lett. **48** pp 272-4.

Hermann J P 1973 Opt. Comm. **9** pp 74-9.

Hopf F A and Stegeman G I 1986 *Applied Classical Electrodynamics Volume 2: Nonlinear Optics* (New York: John Wiley & Sons) p 95.

Horsthuis W H G and Krijnen G J M 1989 to be published.

Lattes A, Haus H A, Leonberger F J and Ippen E P 1983 IEEE J. Quant. Electr. QE-**19** pp 1718-23.

Maker P D and Terhune R W 1965 Phys. Rev. **137** pp A801-18.

Möhlmann G R, Horsthuis W H G, Diemeer M B J, Suyten F M M, Trommel E S, Mc Donach A and Mc Fadyen N 1989 in: *Nonlinear Guided-Wave Phenomena: Physics and Applications* (Houston, Texas) pp 171-4.

Patela S, Jerominek H, Delisle C and Tremblay R 1986 J. Appl. Phys. **60** pp 1591-4.

Rao D N, Swaitkewiewics J, Chopra P, Ghosal S K and Prasad P N 1986 Appl. Phys. Lett. **48** pp 1187-9.

Pump- and probe beam measurements in organic materials

S Graham[1], R Renner[1], C Klingshirn[1], W Schrepp[2], R Reisfeld[3], D Brusilovsky[3] and M Eyal[3]

[1] Fachbereich Physik der Universität, Erwin-Schrödinger-Strasse, D-6750 Kaiserslautern, Federal Republic of Germany

[2] BASF Aktiengesellschaft, Kunststofflaboratorium, D-6700 Ludwigshafen, Federal Republic of Germany

[3] Department of Inorganic and Analytical Chemistry, Hebrew University of Jerusalem, Jerusalem, Israel

[+] Partly supported by the National Council for Research and Development of Israel, by the European Community and by the Materialforschungsschwerpunkt of Rheinland-Pfalz

ABSTRACT: Pump- and probe beam measurements with nanosecond pulses (λ_{exc} = 500 nm) have been performed in different organic materials. In a polydiacetylene Langmuir-Blodgett film we observe a bleaching of the absorption edge, whereas an acridine dye doped glass shows bleaching and induced absorption at the absorption edge. Time-resolved luminescence measurements in the glass yield a strong component in the nanosecond range and a much weaker delayed luminescence component in the millisecond range.

1. INTRODUCTION

Over the past few years considerable research effort has been concentrated on nonlinear optical studies in organic materials, mainly on χ^2 and χ^3 effects (Chemla and Zyss 1987). In our group much experience in pump- and probe beam experiments (Majumder et al 1985, Swoboda et al 1988) as well as in laser-induced gratings (Weber et al 1988) in II-VI semiconductors has been gathered. We have studied the change of the optical properties in semiconductors induced by real or by coherent excitation. We observe with increasing excitation, i.e. with increasing electron-hole pair density, scattering processes, biexciton transitions and finally band gap renormalization and band filling effects. This results for different materials and excitation conditions in a blue or red shift of the absorption edge, in optical gain and in dispersive changes. Our aim is to apply the pump- and probe beam spectroscopy also to organic materials, e.g. in order to compare their behaviour with that of semiconductors.

2. PUMP- AND PROBE BEAM SPECTROSCOPY IN ORGANIC MATERIALS

The pump- and probe beam experiments in polydiacetylene (12/8 diynoic acid) Langmuir-Blodgett (PDA-LB) films have been performed at room temperature. The samples have been excited by nanosecond pulses at λ_{exc} = 500 nm. These quasi-stationary measurements have also been performed in order to complement the time-resolved pump- and probe beam experiments in PTS-PDA done by B.I. Greene et al (1988).

We observe a bleaching of the absorption edge which is situated around 560 nm in a 0.3 μm thick PDA film. At lower excitation intensities (fig. 1), up to 1.7 MW/cm^2, this bleaching is reversible, i.e. the original transmission spectra recover after excitation. At higher excitation intensities, $I_{exc} \geq 3.4$ MW/cm^2, this bleaching is only partly reversible (fig. 2). The solid curve represents the original state before the first excitation, the dashed curve the state during excitation and the dash-dotted curve the state to which the sample returns after excitation. When the sample is excited for a second time the dashed curve represents again the state during excitation, whereas the dash-dotted curve now represents both the state before and after the excitation. So this process is reversible again.

Similar effects of bleaching have also been observed in II-VI semiconductors (Swoboda et al 1988 and in press). In CdS the absorption edge at high temperatures (300 K) is determined by an exponential tail according to the Urbach-Martienssen rule. Under high excitation the filling of these states overcompensates the band gap renormalization due to many-particle effects and thus leads to a blue shift or a bleaching of the absorption edge. State filling could be a possible cause for the reversible part of the bleaching effect in the PDA films, too.

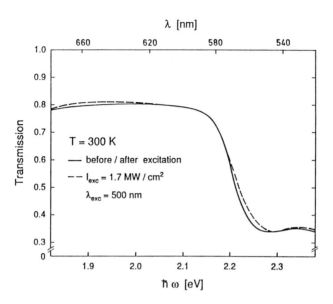

Figure 1: Transmission spectra at 300 K of a PDA-LB film without (-) and during excitation at 500 nm with I_{exc} = 1.7 MW/cm^2 (---)

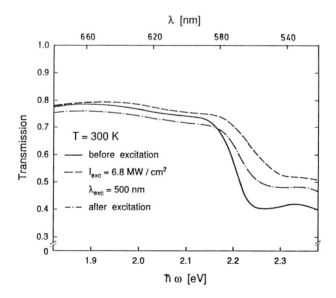

Figure 2: Transmission spectra at 300 K of a PDA-LB film before (-), during excitation at 500 nm with I_{exc} = 6.8 MW/cm^2 (---) and after excitation (-·-)

For the irreversible part of the bleaching in the PDA-LB film several in terpretations are possible. During the fabrication process of the LB film regions with different degrees of polymerisation are formed. Under laser excitation single monomers might be linked to existing polymers leading to a different degree of polymerisation. The reverse effect, that polymers are split into several parts by the laser excitation, is also possible. Another possibility would be a partial melting that is a laser annealing of the sample.

Pump- and probe beam measurements under nanosecond excitation were also performed in an acridine orange dye (AOG) which was embedded in a lead-tin glass (Tick et al 1987 and 1984, Tompkin et al 1987). At excitation intensities of 1.5 MW/cm^2 and more (fig. 3) we see a weak bleaching of the absorption edge around 530 nm and an induced absorption at lower energies. The time needed after the excitation to reach the original state again lies here in the range of minutes.

The effect of induced absorption also occurs in the II-VI semiconductor CdS (Swoboda et al 1988 and in press) at low temperatures (5 K). There the band gap renormalization in an electron-hole plasma by exchange and correlation is more efficient than the state filling. However, the shape of the change of the transmission spectra in CdS is comparable with the one in AOG only in the range between 2.2 and 2.3 eV. The rather flat feature below 2.2 eV might be due to transitions starting from a metastable state which are populated by the excitation. This interpretation could also explain the long time constant.

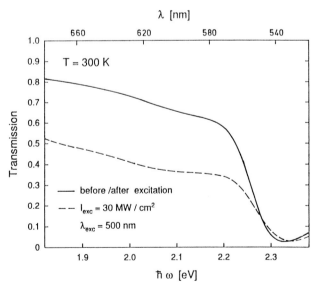

Figure 3: Transmission spectra at 300 K of AOG without (-) and during excitation at 500 nm with I_{exc} = 30 MW/cm^2 (---)

3. LUMINESCENCE MEASUREMENTS IN AN ACRIDINE DYE DOPED GLASS

Luminescence measurements in AOG were performed under high excitation (λ_{exc} = 485 nm) at different temperatures (fig. 4). With decreasing temperatures both main luminescence peaks increase, the one at 625 nm increases more than the other peak at 540 nm.

In time-resolved experiments the decay of the luminescence was measured. We observe a strong component of luminescence in the nanosecond range and a much weaker component in the millisecond range. The latter is nonexponential having an asymptotic lifetime of 27 milliseconds. The luminescence in the nanosecond range is caused by the transition of carriers from the excited singlet state back to the singlet ground state. The possibility of a luminescence on a nanosecond time scale and a corresponding energy diagram have also been mentioned by Tick and Hall (1987).

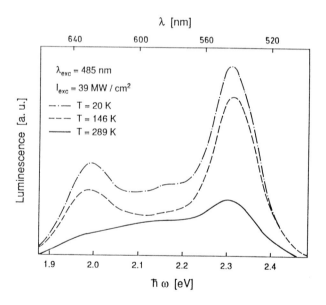

Figure 4: Luminescence of AOG at an excitation intensity of 39 MW/cm^2 (λ_{exc} = 485 nm) at different temperatures.

The luminescence in the millisecond range is a delayed luminescence: after the excitation of carriers from the singlet ground state to the excited state, the carriers relax to the lowest lying triplet state. There they are trapped for several milliseconds before falling back into the singlet ground state. Our results concerning the luminescence in the millisecond range are similar to those of

Tompkin et al (1987). The discrepancy in the decay times of luminescence and induced absorption is not yet clear.

4. CONCLUSIONS AND OUTLOOK

We state that effects like bleaching and induced absorption of the absorption edge caused by real excitation and not by coherent χ^ν effects can be measured by pump- and probe beam spectroscopy in organic materials. First experiments with degenerate laser-induced gratings in the polydiacetylene film and with an organic solution (PYC) have shown diffracted orders of the grating. In both cases the grating is caused by a change in the absorption. Further experiments will give more information on how processes like many-particle effects, diffusion and recombination of carriers cause the optical nonlinearities.

5. REFERENCES

Chemla D S and Zyss J 1987 Nonlinear Optical Properties of Organic Molecules and Crystals (Orlando: Academic Press)

Greene B I, Mueller J F, Orenstein J, Rapkine D H, Schmitt-Rink S and Thakur M 1988 Phys. Rev. Letters **61** 325

Klingshirn C and Haug H 1981 Phys. Reports **70** 315

Majumder F A, Swoboda H E, Kempf K and Klingshirn C 1985 Phys. Rev. **B32** 2407

Swoboda H-E, Majumder F A, Lyssenko V G, Klingshirn C and Banyai L 1988 Z. Phys. **B70** 341

Swoboda H-E, Sence M, Majumder F A, Rinker M, Bigot J-Y, Grun J B and Klingshirn C in press Phys. Rev. B

Tick P A 1984 Phys. and Chem. Glasses **25** 6

Tick P A and Hall D W 1987 Diffusion and Defect Data **53-54** 179

Tompkin W R, Boyd R W, Hall D W and Tick P A 1987 J.Opt.Soc.Am. **B4** 1030

Weber Ch, Becker U, Renner R and Klingshirn C 1988 Z. Physik **B72** 379 and Appl. Phys. **B45** 113

Langmuir-Blodgett films for non-linear optical applications

S Allen

ICI Wilton Materials Research Centre, PO Box 90, Wilton, Middlesbrough,
Cleveland, TS6 8JE.

ABSTRACT: The merits of the Langmuir-Blodgett technique for the
production of ordered thin films having high optical nonlinearities are
discussed in comparison to other methods of film fabrication. The
chemical design of molecular structures to maximise the macroscopic film
nonlinearities   is   discussed. Experimental   techniques   for   the
determination of second-order nonlinear optical coefficients, relating
to second harmonic generation and the linear electro-optic effect are
reviewed and a comparison of the results obtained from different
compounds is given. Finally, the potential of Langmuir-Blodgett films
for the commercial production of nonlinear optical and electro-optic
devices is considered.

1. INTRODUCTION

Much of the pioneering work in the demonstration of the utility of organic
compounds for second-order non-linear optical processes has concentrated on
the use of single-crystalline materials (Zyss and Chemla 1987), and more
recently on polymer-based compounds (Allen 1989b). Although both these
approaches have a number of points in their favour, they also have a number
of disadvantages as outlined below. In principle the use of single crystal
compounds will lead to the highest nonlinearities – the nonlinear optical
molecular unit can be used without the need to add diluting groups such as
aliphatic chains or polymer backbones, and given the right crystal
structure the degree of noncentrosymmetric order can be very high. However
in practice, even in crystals it is often necessary to add bulky
substituent groups in order to induce noncentrosymmetric order. Also the
inability, to date, to predict the structure with which a given molecule
will crystallise is a serious drawback. The process of finding an efficient
nonlinear optical crystalline material remains somewhat hit-and-miss.
Nevertheless, a number of crystalline compounds showing great promise have
been reported over the last few years (Lipscombe et al. 1981, Bailey et al.
1989, Allen 1989a). An additional problem with crystalline compounds
relates to the production of thin films. Many applications of nonlinear
optics, particularly in the field of optical communications, will require
devices in waveguide geometries. In this case it is necessary to fabricate
thin films (a few micrometers in dimension), within which the the molecules
are ordered with respect to each other, and with a given orientation with
respect to the substrate. Although there have been a few reports of
epitaxial growth of organic crystals (Le Moigne 1988), this technique has
not yet been successfully applied to any of the more efficient nonlinear

optical compounds. The most successful approach to waveguide formation in crystalline organic compounds has involved their growth within the cores of thin capillary fibres (Nayar 1988, Vidakovic et al. 1987).

Recently much effort has been concentrated worldwide in the development of nonlinear optical materials based on polymeric compounds. As described in detail elsewhere (de Martino et al. 1988, Lytel et al. 1989, Allen 1989b), this involves the incorporation of a nonlinear optical molecule within a polymer host matrix, either by dissolution or by chemical attachment, their orientation by the application of an electric field whilst heating the polymer, and then cooling the compound down to room temperature so as to freeze in the orientational order. The main advantages of the technique lie in the relatively easy (and cheap) fabrication processes involved, the fact that in principle any nonlinear molecule can be used, and in the good optical quality (in terms of low scattering and absorptive losses) that can be achieved in polymeric materials. The fabrication processes are also well suited to the manufacture of complex waveguide devices. However, the polymeric matrix is inactive, and dilutes the overall nonlinearity - the nonlinear optical molecules will rarely constitute more than a quarter (by weight) of the total. Also the degree of order that can be achieved is limited by thermal motion, and is significantly less than that obtained in an optimised crystal structure. Relaxation of the induced order with time can be a further problem.

An alternative method of fabrication of ordered thin films is that of Langmuir-Blodgett (LB) film deposition. The technique, which has been extensively reviewed (see for example Roberts 1984), can result in the production of thin films of precisely controlled dimensions and structure. In comparison with the materials discussed above, the main advantages of LB films are

- The high degree of order that can be achieved within a single Langmuir monolayer. This can be comparable with that achieved within an optimised crystal structure.

- Their geometry is naturally compatible with waveguide devices. In principle micrometer thick films can be fabricated, although there remain a number of problems in this area. Thinner films can be used as active overlayers on prefabricated waveguides.

- The precise control of thickness that can be achieved by the LB technique makes it well suited to parametric nonlinear optical processes such as second harmonic generation, parametric amplification and frequency mixing. These processes require phase-matching of the waves of different frequency involved in the process, and this can only be achieved in waveguide structures if the film thickness is tightly controlled.

In addition to this is the fact that the LB film (monolayer or multilayer) can provide a useful tool for molecular characterisation. As described in more detail in section 3 below, it can be used as an alternative to the EFISH technique for the determination of molecular hyperpolarisabilities β. As with the other techniques, LB films do have a number of shortcomings:

- Molecules have to be specially designed and synthesised so as to be suitable for LB deposition. This usually requires the incorporation of polar headgroups and aliphatic hydrophobic chains on opposite ends of the molecule. These additional groups do not in general contribute to the molecular nonlinearity β, and so will dilute the macroscopic nonlinear optical coefficient.

- The process of building up films thick enough for waveguiding one monolayer at a time is slow, and therefore expensive on a commercial basis.

- The optical quality of LB films is in general rather poor, with large scattering losses being observed when they are used as waveguides. These losses may be due to the formation of domains of long-range order within the film. It should be possible to reduce this effect by suitable chemical design.

2. DESIGN OF LB MOLECULES FOR HIGH $\chi^{(2)}$ FILMS

As described amply in the literature, the basic requirement for a molecule capable of deposition as a Langmuir monolayer is that it has one end that is hydrophilic, and the other end hydrophobic. Typically the hydrophilic end will consist of a polar headgroup such as -COOH, whilst the the hydrophobic group would be a long aliphatic fatty chain. In order to form a monolayer having a high nonlinearity $\chi^{(2)}$, a unit having a high molecular nonlinearity β must be sandwiched between the hydrophilic and hydrophobic groups. A number of different nonlinear optical units have been investigated, including stilbenes, azobenzenes, merrocyanines, phenylhydrazones and polyenes. Some of the structures are shown in Figure 1. It can be seen that usually the hydrophobic chain has been attached (for synthetic convenience) to the donor group (usually -NRR') of the nonlinear optical unit, although it can be placed at either end.

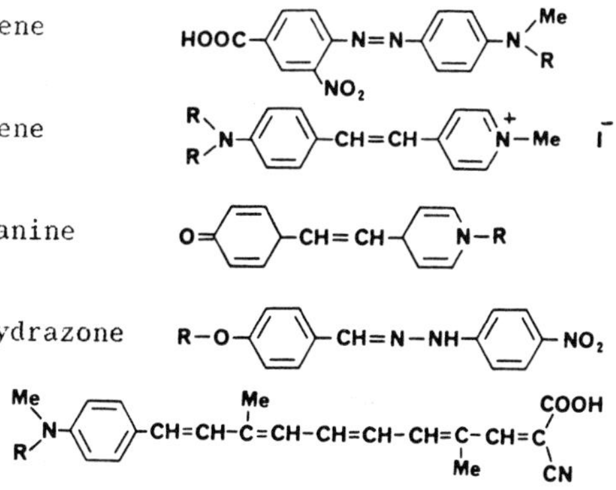

azobenzene — β = 0.12 Ledoux et al (1987)

stilbazene — β = 0.84 Lupo et al (1988)

merrocyanine — β = 1.13 Peterson (1989)

phenylhydrazone — β = 0.17 Lupo et al (1988)

polyene — β = 0.35 Allen et al (1988)

Figure 1: Examples of nonlinear optical LB compounds: β-values are in units of $(nm)^4/V$

The production of a **multilayer** having high $\chi^{(2)}$ is not quite so simple. Usually an LB compound will be deposited as a Y-type film, in which the monolayers deposited on immersion and withdrawal of the substrate from the subphase are oppositely oriented with respect to the substrate, so that the molecular nonlinearities of the two layers cancel each other out resulting in an inactive film. For some molecules, deposition occurs only (or at least preferentially) on immersion, or only on withdrawal. This results in films (called X-type and Z-type respectively) where there is incomplete cancellation of the molecular nonlinearities, and having a finite value of $\chi^{(2)}$. Such films do however tend to be rather unstable with respect to the energetically favoured Y-type arrangement, although high nonlinearities have been measured in a number of Z-type multilayer structures (Aktsipetrov et al. 1985, Ledoux et al. 1987, Hayden et al. 1987). Multilayers with the highest nonlinearities, and greatest stability, can be produced in the "alternating Y-type" configuration (Zyss 1985, Neal et al. 1986). In this case two different molecules are used, as shown in Figure 2. Molecule A is

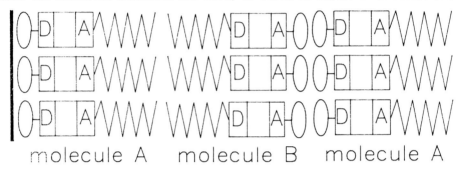

Figure 2. Alternating Y-type deposition of LB films.

deposited on immersion of the substrate into the subphase. The substrate is then passed under a barrier and withdrawn from the subphase through a monolayer of molecule B, which is deposited on top of the monolayer of Molecule A. By repeating this process a multilayer having the structure ABABAB... can be built up. In the simplest case, molecule A might be a nonlinear optical LB molecule, and molecule B an inactive LB molecule such as arachidic acid. Since all the nonlinear optical molecules (A) are now aligned in the same sense, their molecular nonlinearities ß are additive, and despite the fact that molecule B will dilute the macroscopic nonlinearity, a film of high $\chi^{(2)}$ is obtained.

Even higher nonlinearities can be obtained by design of complementary molecules for use in alternating Y-type structure. If molecules A and B are designed so that their nonlinear optical units have the opposite "direction" (as defined, for example, by a vector from the electron donor component of the molecule to the electron acceptor component) with respect to the hydrophilic and hydrophobic end groups (as shown in Figure 2), then all the ß-values will add up in the multilayer resulting in a very high nonlinearity. Examples of molecular structures used in this way are shown in Figure 3.

3. SECOND HARMONIC GENERATION.

Second harmonic generation (SHG) has been the major tool for the characterisation of the nonlinear optical properties of LB films. In addition to the direct measurement of film $\chi^{(2)}$ values (and hence assessment of their potential for use in nonlinear optical devices) the

molecule A	molecule B	$\chi^{(2)}$ (pm/V)	ref.
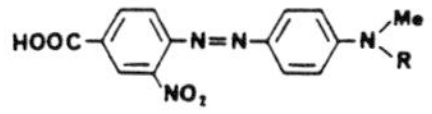 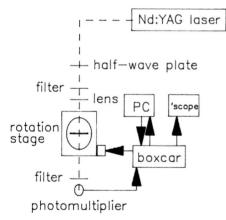	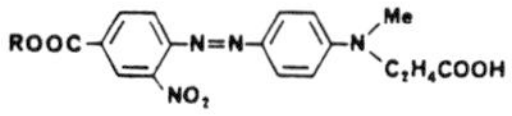	340	Ledoux et al (1988)
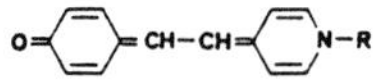	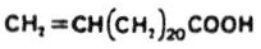	550	Girling et al (1985)
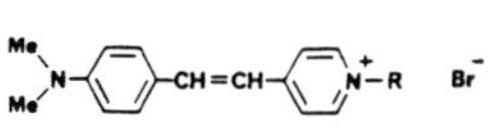	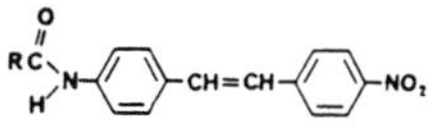	400	Neal et al (1986)
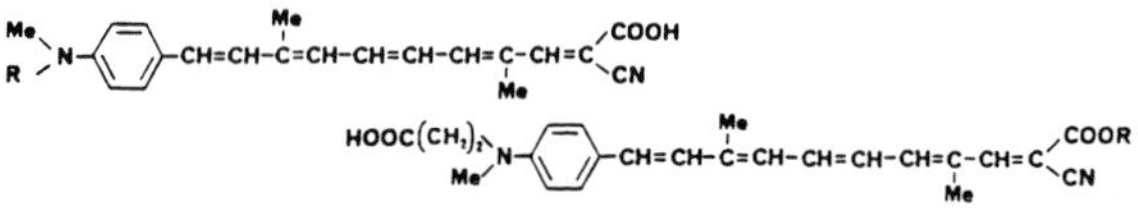			Allen et al (1988)

Figure 3. Alternating Y-type molecules

measurements can be used as an alternative to the well-known EFISH technique (Levine and Bethea 1974) for the determination of molecular ß-coefficients. Although many compounds have by now been characterised using the EFISH technique (Nicoud and Twieg 1987), there are a number of difficulties associated with the experiment and the subsequent analysis. Also it is only possible to evaluate the vector component β_x by this method. As described below, SHG measurements on monolayers or multilayers can be used to derive molecular ß-coefficients, as well as details of the packing of the molecules within the film.

The experimental arrangement for SHG measurements is shown in Figure 4. A Q-switched Nd:YAG laser, operating at 1.064 μm (or 1.91 μm with the incorporation of a H_2 Raman cell) is incident, via appropriate filters, a λ/2 plate for polarisation control and a focusing lens, on the sample. This is mounted on a computer-controlled rotation stage, so that the angle of incidence of the laser beam to the sample can be varied. The second harmonic signal is detected by a photomultiplier tube, after the removal of the fundamental with appropriate filters. The signal from the photomultiplier is averaged with a Boxcar, and recorded on the computer

Figure 4: Experimental apparatus.

as a function of the angle of incidence. Polarising prisms can be inserted both before and after the sample so as to allow a study of all combinations of p- and s- polarised fundamental and harmonic beams.

A typical plot of harmonic intensity as a function of angle of incidence, for a glass substrate coated on both sides with an LB film, is shown in Figure 5. The fringing pattern is derived from interference between the harmonic signals from the films on the front and back faces of the substrate (Allen et al. 1988), and gives no information about the LB film

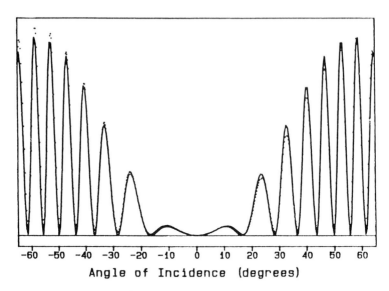

Angle of Incidence (degrees)

Figure 5. Experimental (points) and theoretical (line) SHG intensity

properties. The height and shape of the envelope of the fringe pattern (measured with respect to the second harmonic intensity obtained from a standard sample, usually a polished slice of quartz) lead to the determination of a number of components of the $\chi^{(2)}$ tensor, as detailed below.

The application of standard theories electromagnetic wave propagation to the problem of second harmonic generation within a single thin film, of thickness 1, is given, for example by Kajzar and Messier (1985). It can be shown that the electric field $E^{2\Omega}$ at the second harmonic frequency output from the film is given by the expression

$$E^{2\Omega} = \frac{4\pi\chi^{(2)}(\theta)}{\delta\epsilon} e^{i\Phi_{2\Omega}} [a_1(e^{i\Phi_+}-1)+a_2(e^{i\Phi_-}-1)]/[a_3-a_4e^{i2\Phi_{2\Omega}}] \qquad (1)$$

where
$$\Phi_\Omega = 2n_\Omega(1\Omega/c)\cos\theta_\Omega$$
$$\Phi_{2\Omega} = 2n_{2\Omega}(1\Omega/c)\cos\theta_{2\Omega}$$
$$\Phi_\pm = \Phi_{2\Omega} \pm \Phi_\Omega$$
$$\delta\epsilon = (n_\Omega)^2 - (n_{2\Omega})^2$$

and the coefficients a_i are functions of the refractive indices $n_{2\Omega}$ and n_Ω of the film, the refractive indices of the surrounding media and the angle of incidence of the laser beam. The form of the angle dependent nonlinear optical coefficient $\chi^{(2)}(\theta)$ depends on the polarisation of the fundamental and harmonic beams. The above equation can be simplified somewhat if the LB films are sufficiently thin (Allen et al. 1988).

In the configuration described above second harmonic generation occurs within films on either side of the substrate, and the output electric field is the sum of the fields from the two films, with the appropriate phase factors due to propagation through the glass substrate.

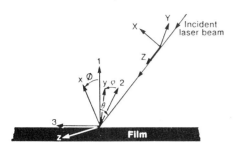

Figure 6: Coordinate systems.

The equations can be used directly to evaluate the components of the nonlinear optical coefficient $\chi^{(2)}$ of the Langmuir Blodgett film. The coordinate system used is shown in Figure 6. Here the fundamental laser beam is incident at an angle θ to the film normal. We can define a "beam" set of coordinates (X,Y,Z) such that Z lies along the beam direction, X lies within the plane of incidence and Y perpendicular to the plane of incidence. In this case, p-polarised and s-polarised beams have electric field components E_X and E_Y respectively. A "film" set of coordinates $(1,2,3)$ is defined with 1 along the film normal and 3 and 2 in and perpendicular to the plane of incidence respectively. The angle dependent nonlinear optical coefficient "seen" by the incident beam will be χ_{XXX} when both incident and harmonic beams are p-polarised, χ_{XYY} for s-polarised fundamental generating p-polarised harmonic and so on. These coefficients are obtained in terms of the film nonlinearities χ_{ijk} $(i,j,k = 1,2,3)$ by using the standard equations for transformation of tensor properties, giving

$$\chi_{XXX} = \sin^3\theta\chi_{111} + 2\sin^2\theta\cos\theta\chi_{113} + \sin\theta\cos^2\theta\chi_{133} + \sin^2\theta\cos\theta\chi_{311}$$
$$+ 2\cos^2\theta\sin\theta\chi_{313} + \cos^3\theta\chi_{333} \qquad (2)$$

If we make the simplifying assumption that the Langmuir Blodgett film is uniaxial in symmetry, with the axis of symmetry being along the film normal, then this reduces to

$$\chi_{XXX} = \sin^3\theta\chi_{111} + \sin\theta\cos^2\theta\chi_{133} + 2\cos^2\theta\sin\theta\chi_{313} \qquad (3)$$

with similar equations for the other components.

The relative magnitudes of the components χ_{111}, χ_{133} and χ_{313} determine the precise shape of the envelope of the second harmonic output pattern, and so the magnitude all three components can in principle be obtained by fitting the experimental results to this theory. An example of the high level of agreement between theory and experiment that can be achieved was shown in Figure 5. By utilising different polarisation geometries in the experiment it is possible to determine most of the independent coefficients of the films nonlinear optical tensor $\chi^{(2)}$.

Having determined the magnitude of the different $\chi^{(2)}$ coefficients, we can then derive useful information relating to the orientation of the molecular species within the film. We can define a set of "molecular axes" (x,y,z), also shown in Figure 6, with x corresponding to the long axis of the molecule. In general the long axis of the molecule will be inclined at an angle ϕ to the film normal, and the axis y, defined to lie within the film plane, will lie at some azimuthal angle φ with respect to the film axis 2. The (oversimplified) description that we use of the molecular alignment within the Langmuir Blodgett film is that all molecules have the same "tilt angle" ϕ, but that the azimuthal angles vary from molecule to molecule. The average molecular polarisability $\langle\beta\rangle$ in the film frame of reference is found by averaging over all molecular orientations, again using standard transformation equations. The simplest average to use is that all azimuthal angles are equally probable, although there is evidence that in practice the molecules align preferentially with the dipping direction.

Using this approximation results in the following expression for $\langle\beta_{111}\rangle$:

$$\langle\beta_{111}\rangle = \cos^3\phi\,\beta_{xxx} - 2\cos^2\phi\sin\phi\,\beta_{xxz} + \cos\phi\sin^2\phi\,\beta_{xzz} - \sin\phi\cos^2\phi\,\beta_{zxx}$$

$$+ 2\cos\phi\sin^2\phi\,\beta_{zxz} - \sin^3\phi\,\beta_{zzz} \qquad (4)$$

with similar expression for the other components of $\langle\beta\rangle$.

This expression can be much simplified if we make the rather drastic assumption that the only significant molecular β component is associated with the long axis of the molecule, so that $\beta_{xxx} = \beta$, and all other components are zero. In this case,

$$\langle\beta_{111}\rangle = \cos^3\phi\,\beta; \quad \langle\beta_{133}\rangle = \cos\phi\sin^2\phi\,\beta/2 = \langle\beta_{313}\rangle \qquad (5)$$

The film nonlinearities $\chi^{(2)}$ described earlier can be written in terms of these averaged molecular polarisabilities using the standard equations

$$\chi_{IJK} = N\,f^{2\Omega}(f^\Omega)^2\langle\beta_{IJK}\rangle \qquad (6)$$

so that, for example, the angular dependent effective nonlinearity χ_{xxx} can be written in terms of the molecular nonlinearity β and the tilt angle ϕ as

$$\chi_{xxx} = N\,f^{2\Omega}(f^\Omega)^2[\sin^3\theta\cos^3\phi + 1.5\cos\phi\sin^2\phi\sin\theta\cos^2\theta]\beta \qquad (7)$$

The effect of variations in the molecular tilt angle ϕ on the shape of the envelope of the second harmonic intensity against angle of incidence plot is shown in Figure 7. As the tilt angle of the molecule increases, the maximum of the envelope function moves in to smaller angles of incidence, and the half-width of the envelope function increases. These two easily measured quantities can be used to determine ϕ within this model.

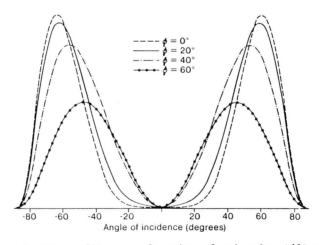

Figure 7. SHG profile as a function of molecular tilt angle ϕ.

This technique has been used to determine the molecular nonlinearities of the compounds shown in Figure 2. The β-values obtained by this technique are in good agreement with those obtained by EFISH measurements or by calculation. However caution is needed when using monolayer SHG measurements to evaluate the merits of an LB compound for device fabrication. Multilayer studies of a number of Z- and alternating Y-type compounds (Peterson 1989, Ledoux et al. 1986) show that the film nonlinearity $\chi^{(2)}$ falls off with increasing number of layers. Thus the intensity of second harmonic does not show the expected square-law dependence on the number of layers in the film. The origin of this decrease in nonlinearity is not clear. It may possibly be due to some interaction between the nonlinear optical molecule and the substrate itself, which can either have the effect of increasing the degree of order in the monolayers close to the substrate, or alternatively may actually enhance the molecular nonlinearity of these layers. Fortunately, the $\chi^{(2)}$ value does not continue to fall, but levels off to a stable value by about 10 monolayers thickness. The thick-film value for $\chi^{(2)}$ can be as much as an order of magnitude less than the monolayer value. Ledoux et al. (1989) have studied this effect by spacing a single active monolayer of an azobenzene chromophore from a glass substrate by a variable number N of inactive "buffer" layers. It was shown that the enhancement of the nonlinearity obtained for N=0 is greatest when the molecule is deposited on a hydrophilic substrate. In this case the nonlinear optical unit of the LB molecule is adjacent to the substrate. For hydrophobic substrates, when the nonlinear optical unit is spaced from the substrate by the fatty chain, the enhancement is reduced. These observations have been repeated with other molecules (Allen et al. 1988).

As mentioned earlier, the most stable structures are formed by alternating Y-type deposition, and a number of combinations of molecules forming such multilayers have been studied. The results are summarised in Figure 3. It can be seen that very high nonlinearities have been obtained, although in many cases the measurements have been made using a wavelength (1.06 μm) for which the second harmonic (532 nm) is close to resonance with the electronic transitions of the molecule, giving an enhancement $\chi^{(2)}$. Nonlinear optical coefficients an order of magnitude greater than those of lithium niobate ($\chi^{(2)} \approx 60$ pm/V) are reported.

One of the major problems remaining in the implementation of nonlinear optical waveguide devices from LB films is the <u>optical quality</u> of the films, particularly with respect to the optical scattering losses observed in such films (Barnes and Sambles 1987). This may possibly be due to the formation of domains of long range orientational order, as were observed by Peterson et al. (1988) in films of fatty acids. Tredgold et al. (1987) demonstrated that lower scattering losses can be achieved in LB multilayers of preformed polymers. This work was extended (Tredgold et al. 1988, Young et al. 1989) to produce noncentrosymmetric alternating Y-type layers, one component of which was a preformed polymer, whilst the other was an active nonlinear optical merrocyanine molecule. Films up to 160 monolayers thick were deposited, and although the SHG signal again did not show a quadratic dependence on the number of layers, large nonlinearities were observed, and the conversion ratio ($I^{2\Omega}/[I^{\Omega}]^2$) obtained was the highest reported to date for an LB film. Further chemical effort in this area should reduce the optical losses to acceptable values.

4. ELECTRO-OPTIC (POCKELS) EFFECT

Many waveguide applications of nonlinear optical materials will utilise the electro-optic (Pockels) effect rather than second harmonic generation. Examples of electro-optic devices include modulators and directional couplers. In principle the electro-optic coefficients for an LB film can be deduced from a knowledge of the SHG coefficients, and the refractive indices and dielectric constants of the material at all wavelengths of interest. In practice it is more satisfactory to measure the coefficients directly. Cross et al. (1986) have described a technique for doing this that involves the observation of surface plasmon modes in the attenuated total reflection (ATR) spectrum from a thin silver film coated with the LB film. The experimental apparatus is shown in Figure 8. The surface of a

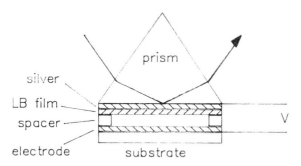

Figure 8. Sample configuration for ATR electro-optic measurements

glass prism is coated with a silver film about 55 nm thick. When the internally reflected intensity from this prism/silver/air interface is monitored as a function of the angle of incidence, a sharp minimum is observed at one particular angle, as shown in Figure 9a, due to the coupling of energy from the beam into a surface plasmon mode of the silver film (Swalen 1979). The position, and shape of this minimum is very

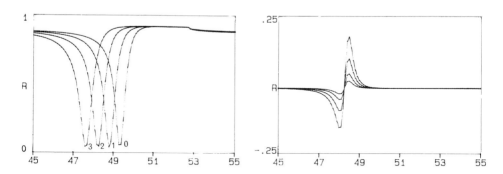

Figure 9. (a) Shape of ATR curve for bare silver, and silver coated with 1, 2 and 3 LB monolayers. (b) Change in reflectivity (%) from an LB bilayer expected for changes in film refractive index of 1,2,3 and 4 x 10^{-4}.

sensitive to changes in the refractive index of the media surrounding the silver film. If the silver is coated with a Langmuir monolayer, there will be a significant shift in the position of the minimum. This shift has been used to determine the refractive index and thickness of the deposited film (Pockrand et al. 1977). Using the electrode geometry shown in Figure 8, an alternating electric field can be applied across the LB film. This will change the refractive index of the film, via the Pockels effect, and this change will modify the shape of the minimum in the ATR spectrum. The change in reflectivity is measured using a lock-in amplifier referenced to the applied voltage, and has the from shown in Figure 9b. Analysis of the signal as a function of the angle of incidence will lead directly to the change in refractive index induced by the applied field, and hence to the electro-optic r-coefficients for the film. A study of the electro-optic effect in alternating Y-type multilayers (Cross et al., 1987) found large imaginary parts of nonlinear optical coefficient, which are assumed to be due to field dependent scattering effects.

To operate a real device based on the above concepts it would be necessary to produce a much greater modulation depth than was reported in these experiments. This could possibly be achieved by either using much thicker LB multilayers, or increasing significantly the voltage applied across the film (or both!).

5. CONCLUSIONS

The design and characterisation of LB films for use in second-order nonlinear optical applications has been reviewed. It should be clear from the above that the LB technique is capable of producing thin films having nonlinear optical coefficients as good as, or better than any of the competitive technologies. The method is proven as a reliable method for the assessment of molecular nonlinearities β. The step-up from thin (i.e. < 100 nm) films, for characterisation in through-film geometries, to micrometer scale films capable of supporting low-loss propagation of waveguide modes is considerable. The two major hurdles lie in the improvement of the optical quality of the films and in the adaption of the technique from laboratory to commercial scale. The first of these requires a chemical and materials science solution. Routes described above, involving the use of polymer chemistry, would seem to offer the best hope. The second problem is one of engineering, and given the impetus that would result from solution of the first problem, should be resolvable. Thus although the future for LB films in the real world of nonlinear optical device manufacture is still at best uncertain their potential is great.

ACKNOWLEDGEMENTS

Work on LB films in ICI has involved collaboration between a number of chemists and physicists. In particular I acknowledge the contributions of Paul Gordon, Mike Hutchings and Brian Bothwell to the chemistry, and Tom McLean, Ron Swart and Sarah Frogett for their LB film expertise. Funding from the EEC through their ESPRIT program is also acknowledged.

REFERENCES

Aktsipetrov O A, Akhmediev N N, Baranova I M, Mishina E D and Novak V R 1985 Sov. Phys. JETP **62** 524.
Allen S 1989a Organic Materials for Non-linear Optics ed Hann R A and Bloor D (London: RSC) pp 137-150.
Allen S 1989b Proc. SPIE **1120** in press.
Allen S, McLean T D, Gordon P F, Bothwell B D, Robin P, and Ledoux I 1988 Proc. SPIE **971** 206.
Bailey R T, Cruikshank F R, Guthrie S M G, McArdle B J, McGillivray G W, Morrison H, Pugh D, Shepherd E A, Sherwood J N, Yoon C S, Kashyap R, Nayar B K and White K I 1989 Organic Materials for Nonlinear Optics ed Hann R A and Bloor D (London: RSC) pp 129-136.
Barnes W L and Sambles J R 1987 J. Phys D **20** 1125.

Cross G H, Girling I R, Peterson I R and Cade N A 1986 Elect. Lett. **22** 1111.

Cross G H, Peterson I R and Girling I R 1987 Proc. SPIE **824** 79.

DeMartino R, Haas D, Khanarian G, Leslie T, Man H T, Riggs J, Sansone M, Stamatoff J, Teng C and Yoon H 1988 Nonlinear Optical Properties of Polymers ed Heeger A J et al. (MRS Symposium Proc. **109**, Pittsburgh: MRS) pp 65-76.

Girling I R, Kolinsky P V, Cade N A, Earls J D and Peterson I R 1985 Opt. Comm. **55** 289.

Hayden L M, Kowel S T and Srinivasan M P 1987 Opt. Comm **61** 351.

Kajzar F and Messier J 1985 Phys. Rev. A **32** 2352.

Ledoux I, Josse D, Vidakovic P, Zyss J, Hann R A, Gordon P F, Bothwell B D, Gupta S K, Allen S, Robin P, Chastaing E and Dubois J-C 1987 Europhys. Lett. **3** 803.

Ledoux I, Josse D, Fremaux P, Piel J P, Post G, Zyss J, McLean T D, Hann R A, Gordon P F and Allen S 1988 Thin Solid Films **160** 217.

Ledoux I, Josse D, Zyss J, McLean T, Hann R A, Gordon P F, Allen S, Lupo D, Prass W, Scheunemann U, Laschewsky A and Ringsdorf H 1989 Nonlinear Optical Effects in Organic Polymers ed Messier J et al. (Kluwer Academic Publishers) pp 79-91.

Le Moigne J, Thierry A, Chollet P A, Kajzar F and Messier J 1988 Proc. SPIE **971** 173.

Levine B F and Bethea C G 1974 Appl. Phys. Lett. **24** 445.

Lipscombe G F, Garito A F and Narang R S 1981 J. Chem. Phys. **75** 1509.

Lupo D, Prass W, Scheunemann U, Laschewasky A, Ringsdorf H and Ledoux I 1988 J. Opt. Soc. Am. B **5** 300.

Lytel R, Lipscombe G F, Stiller M, Thackara J I and Ticknor A J 1989 Organic Materials for Non-linear Optics ed Hann R A and Bloor D (London: RSC) pp 382-389.

Nayar B K, White K I, Holdcroft G and Sherwood J N 1988 Nonlinear Optical and Electroactive Polymers ed Prasad P N and Ulrich D R (New York: Plenum) pp 427-437.

Neal D B, Petty M C, Roberts G G, Ahmad M M, Feast W J, Girling I R, Cade N A, Kolinsky P V and Peterson I R 1986 Elect. Lett. **22** 461.

Nicoud J F and Twieg R J 1987 Nonlinear Optical Properties of Organic Molecules and Crystals Volume 2 ed Chemla D S and Zyss J (Orlando: Academic) pp 255-267.

Peterson I R 1989 Organic Materials for Non-linear Optics ed Hann R A and Bloor D (London: RSC) pp 317-333.

Peterson I R, Earls J D, Girling I R and Barnes W L 1988 J. Phys. D **21** 773.

Pockrand I, Swalen J D, Gordon J G and Philpott M R 1977 Surf. Sci. **74** 237.

Roberts G G 1984 Contemp. Phys. **2** 109.

Swalen J D 1979 J. Phys. Chem. **83** 1438.

Tredgold R H, Young M C J, Hodge P and Khoshdel E 1987 Thin Solid Films **151** 441.

Tredgold R H, Young M C J, Jones R, Hodge P, Kolinsky P and Jones R J 1988 Elect. Lett. **24** 308.

Vidakovic P V, Coquillay M and Salin F 1987 J. Opt. Soc. Am. B **4** 998.

Young M C J, Tredgold R H and Hodge P 1989 Organic Materials for Non-linear Optics ed Hann R A and Bloor D (London: RSC) pp 354-360.

Zyss J 1985 J. Molecular Electronics **1** 25.

Zyss J and Chemla D S 1987 Nonlinear Optical Properties of Organic Molecules and Crystals Volume 1 ed Chemla D S and Zyss J (Orlando: Academic) pp 23-191.

Inst. Phys. Conf. Ser. No 103: Section 2.3
Paper presented at Int. Conf. Materials for Non-linear and Electro-optics, Cambridge, 1989

175

Synthesis and LB-monolayer properties of amphiphilic rigid rod molecules

J Nordmann and W Herbst

Siemens AG, P. O. Box 3220, D-8520 Erlangen

ABSTRACT: The amphiphilic rigid rod molecules were
synthesized by reaction of stearoylchloride with -OH and
-NH₂ functionalized rigid rod stilbenes and azobenzenes.
Some of them are liquid crystalline which is confirmed by
thermal analysis and polarizing microscopy. The surface
pressure/area isotherms of the pure compounds and mixtures
with stearic acid were determined. Especially the unexpec-
ted behaviour of 4-stearicacidester-4'-nitrostilbene is
discussed.

1. INTRODUCTION

One way of producing oriented structures for nonlinear-optical
investigations is the Langmuir-Blodgett technique.

In order to process organic compounds to thin films with the
LB-technique the active parts of the molecules have to be substituted with amphiphilic substituents.

Potential candidates for the preparation of organic nonlinear
wave-guide structures possess extended conjugated π-systems
with electron-donating and electron-accepting groups attached
at opposite sides of the molecule. Stilbene- and azobenzene
derivatives with nitro- or cyanosubstituents as electron-
accepting and amine- or oxygenfunctionalities as electron-
donating groups are commonly used (Chemla et al. 1987).

In this study stearic acid derivatives with the structural
elements above described were synthesized and the influence
of these structural variations on the LB-monolayer properties are discussed.

2. EXPERIMENTAL

All reagents are commercially available and were used as
received unless otherwise specified.

Monolayer studies were performed using a Langmuir trough
(Lauda film balance). The materials were spread from an ap-

proximately 1×10^{-3} M solution in chloroform onto a water-subphase at the described temperatures. The water was purified by reverse osmosis, followed by deionization through a Seralpur PRO 90 C system ($\sigma = 0,05$ µS/cm).

3. SYNTHESIS

Scheme 1 shows the synthetic route used for the preparation of the stearic acid derivatives 3a-h. They are obtained by reaction of stearoylchloride with the appropriate $-NH_2$ or $-OH$ functionalized stilbenes and azobenzenes derivatives.

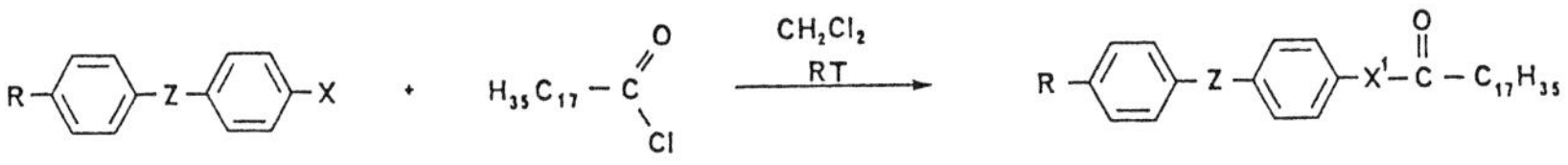

1 + 3	R	Z	X	X'
a	NO_2	N = N	OH	O
b	CN	N = N	OH	O
c	OCH_3	N = N	OH	O
d	NO_2	CH=CH	OH	O
e	OCH_3	CH=CH	OH	O
f	NO_2	CH=CH	NH_2	NH
g	CN	CH=CH	NH_2	NH
h	OCH_3	CH=CH	NH_2	NH

Scheme 1: Synthesis of amphiphilic stearic acid derivatives 3a-h

Some of the characteristic data of compound 3a-h obtained by thermal analysis, polarizing microscopy and UV-spectroscopy are shown in table 1. With the exception of 3c and 3d the compounds are liquid crystalline.

Further details of the reaction conditions and spectroscopic data of 3a-h, obtained by IR and ¹H-NMR-spectroscopy and the preparation of the starting materials 1a-h will be published (Nordmann 1989).

Compound	Phase transition	UV (CH_2Cl_2) λ_{max} (nm)
3a	k 76 s 93 i	360
3b	k 97 s 104 i	331
3c	k 92 i	348
3d	k 84 i	356
3e	k 125 n 131 i	322
3f	k 141 s 159 i	375
3g	k 144 s 159 i	340
3h	k 190 lc 196 i	333

Table 1: Physical data of stearic acid derivatives 3a-h

4. LB-MONOLAYER PROPERTIES

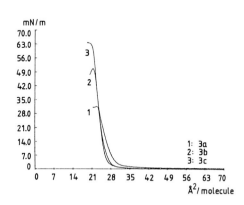

Fig. 1.: Surface pressure area diagram of 3a-c

Figure 1 shows the surface pressure - molecular area isotherms of compounds 3a-c at 20.2°C. The extrapolated area per molecule at zero pressure was estimated to be 25 Å² in all three cases. The different chemical structure of the compounds has no influence on the required area of the molecules but does so on the film collapse pressure. It increases with decreasing polarisability of the molecules, due to the intermolecular forces of the molecules.

The isotherms of compounds 3d and 3e are shown in Figure 2 and 3 for different temperatures.

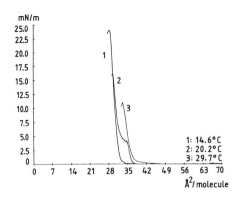

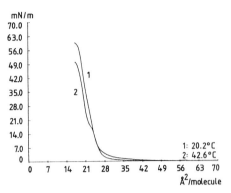

Fig. 2: Surface pressure area isotherms of 3d at various temperatures

Fig. 3: Surface pressure area isotherms of 3e at various temperatures

Two regions are apparent at a temperature of 20.2°C for compound 3d. There seems to be a phase transition at a pressure of 4,8 mN/m, which can be understood by an orientation of the molecules (Berkovic et al. 1987). At lower temperature (14.6°C) and higher temperature (29.7°C) of the subphase the transition disappears. A similar behaviour is known from thermotropic liquid crystals, having phase transitions as a function of temperature.

The isotherms of compound 3e at 20.2°C and 42.6°C (Figure 3) show a curve shape similar to those of compound 3d but the curves are shifted to lower areas per molecule. This effect is reproducible but cannot be explained at the moment.

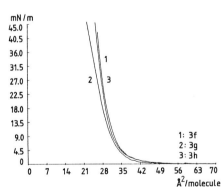

Figure 4: Surface pressure area diagrams of 3f-h

Figure 4 shonws the isotherms of compounds 3f-h. The film forming properties of 3f agree very well with data known from the literature (Ahmad 1986). The area per molecule is somewhat larger than for 3a-c.

Mixtures of 3d with stearic acid in ratios of 9:1 and 1:9 were prepared to understand the behaviour of a phase transition of 3d (figure 5). The monolayer properties are affected by the amount of stearic acid in the mixtures. With an increasing amount of stearic acid

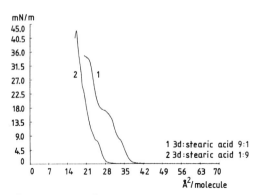

Fig. 5: Surface pressure area diagramms for the mixtures of stearic acid with 3d

the collapse pressure reaches the value of pure stearic acid. But it seems that the compounds are immiscible because the step which is charac-teristic for compound 3d is still present. If the compounds are miscible the step should disappear or change its position.

5. CONCLUSIONS

Some new stearic acid derivatives substituted with rigid rod moieties were synthesized and characterised by thermal analy-sis, polarizing microscopy and UV-spectroscopy. Due to the shape of the molecules some of them are liquid crystalline.

Compounds 3a-c and 3f-h showed the expected behaviour of the isotherms. The different behaviour of 4-stearicacidester-4'-nitrostilbene 3d can be explained by an oriented phase oc-curing between the fluid and the solid phase of the compounds induced by compression and orientation of the molecules. Mix-tures of 3d with stearic acid showed an immiscibility of the compounds at the measuring temperature.

We like to thank Prof. Möhwald (University of Mainz) and his coworkers for discussions and A Maltenberger and N Lalyre for synthetic work.

6. REFERENCES

Ahmad M M, Feast W J, Neal D B, Petty M C and Roberts G G 1986 J. Mol. Electron. 2 129
Berkovic G, Rasing T und Shen Y R 1987 SPIE **824** 115
Chemla D and Zyss J ed 1987 Nonlinear Optical Properties of Organic Molecules and Crystals Vol 1 and 2 (New York: Acade-mic Press)
Nordmann J 1989 submitted to Synthesis

Inst. Phys. Conf. Ser. No 103: Section 2.3
Paper presented at Int. Conf. Materials for Non-linear and Electro-optics, Cambridge, 1989

181

Nonlinear optical Y-type Langmuir-Blodgett films of 2-docosylamino-5-nitropyridine

Ch. Bosshard, B. Tieke*, M. Seifert+, and P. Günter

Institute of Quantum Electronics, ETH Hönggerberg, CH-8093 Zürich, Switzerland
* Ciba Geigy AG, Research Center Plastics and Additives, CH-1701 Fribourg, Switzerland
+ Polymetron AG, CH-8617, Mönchaltorf, Switzerland

ABSTRACT: We have measured linear and nonlinear optical properties of well ordered Y-type Langmuir-Blodgett (LB) multilayers of 2-docosylamino-5-nitropyridine (DCANP). The layers have a good optical quality over the whole film area. They are transparent in the visible and the near infrared. The birefringence and nonlinear optical susceptibilities have been investigated. The anisotropy of the second-harmonic efficiency indicates that the chromophores are aligned in a plane parallel to the dipping direction. In first waveguiding experiments carried out TE_0/TM_0 modes propagating over more than 15 mm could be excited.

1. INTRODUCTION

Organic materials with very interesting nonlinear optical properties have been extensively studied in the past 15 years. Their high quadratic nonlinearities, their sufficient transparency range from the visible to the near infrared spectrum and their short electronic response time to optical excitation make them attractive for all-optical signal processing. Especially suited for these applications are materials in the form of single crystal optical fibers and thin film layers (Zyss 1985). The Langmuir-Blodgett (LB) deposition technique allows the fabrication of thin films with an exactly defined molecular arrangement and film thickness (Kuhn *et al* 1972).

We determined linear and nonlinear optical properties of well-ordered LB-films of 2-docosylamino-5-nitropyridine (DCANP) which does not exhibit a head-to-tail arrangement of the molecules (Y-type structure) (Decher *et al* 1988). DCANP can be described as a long-chain analogue of 2-cyclooctylamino-5-nitropyridine (COANP) which exhibits large nonlinear optical susceptibilities in the crystalline state (Günter *et al* 1987, Bosshard *et al* 1989).

Monolayers of DCANP can easily be transferred onto various hydrophobic substrates (e.g. glass, quartz glass or silicon hydrophobized with octadecyltrichlorosilane, or zinc selenide). High quality films of at least 1.2 µm thickness (270 bilayers) showing no visible turbidity have been prepared by us recently. FTIR-ATR-Spectra (Fourier Transform Infrared Attenuated Total Reflection) of LB-films on zinc selenide are nearly identical with those of bulk material indicating identical chemical compositions.

The bilayer spacing $d_{010} = (4.42 \pm 0.03)$ nm of DCANP was obtained through small angle X-ray scattering (SAXS). Molecular models indicate a maximum length of 3.2 nm for DCANP in the fully extended conformation. This value is 17% shorter than the observed layer spacing. Therefore a head-to-tail arrangement of the molecules (X- or Z-type) can be ruled out. Consequently LB-films of DCANP should be Y-type bilayers with an unusually large tilt-angle of the alkyl-chains of approximately $(50 \pm 10)^{\circ}$ with respect to the surface normal.

2. LINEAR OPTICAL PROPERTIES

The thin film layers of DCANP are of good optical quality (areas of complete extinction between crossed polarizers up to 1 cm^2). Polarized UV/VIS-transmission spectroscopy shows that the chromophores are homogeneously oriented parallel to the dipping direction over the whole film area. The absorption coefficient parallel to the dipping direction $\alpha_{||}$ and the ratio $\alpha_{||}/\alpha_{\perp}$ were determined at the point of maximum absorption ($\lambda = 375$ nm) yielding $\alpha_{||} = 127\,000$ cm^{-1} and $\alpha_{||}/\alpha_{\perp} = 1.6$. The transparency extends from 400 nm to at least 2000 nm. In agreement with these measurements we define the following coordinate system: X_3 is the dipping direction. X_1 is the direction perpendicular to X_3, but parallel to the glass substrate. X_2 is the axis perpendicular to both X_1 and X_3.

The birefringence $\Delta n = n_1 - n_3$ of our LB layers was determined by using a polarizing microscope. As is described by Sutter *et al* (1988) measurements of the birefringence of organic materials with an Ehringhaus compensator are complicated because of strong dispersion. The measurements on our thin film layers which were in the order of 10 to 500 nm were performed with a Senarmont compensator allowing the determination of optical path differences up to one wavelength. The observed birefringence varies from $\Delta n = 0.078$ ($\lambda = 446$ nm) to $\Delta n = 0.049$ ($\lambda = 624$ nm). The dispersion is plotted in Fig.1.

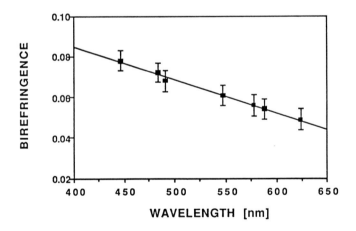

Fig. 1.   Dispersion of the birefringence $\Delta n = n_1 - n_3$ of DCANP.

3. NONLINEAR OPTICAL PROPERTIES

Second-harmonic generation experiments were performed on acentric DCANP LB-layers deposited on hydrophobic fused quartz and pyrex glass substrates. The experiments were carried out with thin films consisting of 5 to 270 bilayers. A Q-switched Nd-YAG laser (Quantronix 416, $P^{\omega} = 2.3$ kW, repetition rate 500 Hz, pulse width 350 ns) operating at $\lambda = 1064$ nm was the light source. The second-harmonic generation efficiency for a light propagation direction perpendicular to the thin film surface is given by (Armstrong *et a*

$$\eta = \frac{2\omega^2}{\varepsilon_o \, c^3 \, (n^\omega)^2 n^{2\omega}} \, d_{eff}^2 \, L^2 \, I^\omega \, \text{sinc}^2 \left(\frac{\Delta k \, L}{2}\right) \tag{1}$$

1962) where $\text{sinc}(x) = (\sin x)/x$, L is the film thickness, d_{eff} is the effective nonlinear optical susceptibility, and $\Delta k = k_2 - 2k_1$ is the phase mismatch between the fundamental and second-harmonic waves with wave vectors k_1 and k_2, respectively. A peak conversion efficiency $\eta = I^{2\omega}/I^\omega = 2 * 10^{-7}$ was observed for $L = 1.2 \, \mu m$ and an intensity $I^\omega = 220$ MW/cm^2 incident on the sample. Using equation (1) and taking into account the reflection losses we can approximate d_{333} from the conversion efficiency η. We obtain a value of $d_{333} = 3.5$ pm/V ($n_{DCANP} = 1.51$). The accuracy of this value should, however, not be overestimated since it has been calculated by using a series of parameters of moderate accuracy.

By varying the polarization direction of the fundamental wave from -90° to 90° (with respect to the X_3-direction) and keeping the analyzer position along the dipping direction we observed a strong anisotropy of the second-harmonic generation efficiency. Fig.2 shows this dependence for 18 bilayers. No second-harmonic wave polarized parallel to X_1 could be detected, indicating that $d_{111} = 0$. The nonlinear optical coefficient d_{333} (in our notation)

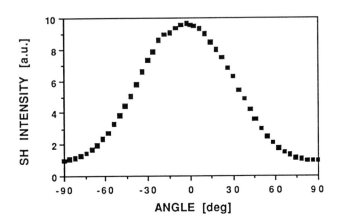

Fig 2.   Dependence of second-harmonic intensity on the polarization of the incident fundamental beam.

was determined by measuring the second-harmonic intensity of a rotated LB-film (around the X_3-axis) with the polarization of E^ω and $E^{2\omega}$ both parallel to X_3. With the polarization of E^ω parallel to X_1 we obtained d_{311}. The Maker-fringes of a 2 mm thick Quartz sample ($d_{111} = 0.4$ pm/V) served as the reference. The nonlinear optical susceptibilities determined from smoothed curves as the one shown in Fig.3 yielded $d_{333} = 6.8 \pm 1.2$ pm/V and $d_{311} = 0.9 \pm 0.2$ pm/V.

For these experiments DCANP samples of different thicknesses were used. Special pyrex substrates with a surface finish of $\lambda/10$ ($\lambda = 633$ nm) and an especially small wedge angle between 5 and 10 angle seconds were used in order to exclude phase differences resulting from an asymmetric substrate. The measurements were performed on samples deposited either on a single or on both sides of the substrate leading to the well-known interference effects between the second-harmonic waves generated within the two films in the latter case. Since the pyrex substrate and DCANP have almost the same refractive indices ($n_{pyrex}(633\text{ nm}) = 1.471$, $n_{DCANP}(633\text{ nm}) = 1.51$) we neglected the Fresnel reflections at the DCANP-glass interfaces.

In the case that both sides of the substrate are covered with DCANP the intensity of the frequency-doubled wave is given by (Shen 1984)

$$I^{2\omega} = \frac{4\omega^2}{\varepsilon_0 c^3 \left(n^{2\omega}\right)^2} d^2 L^2 \left(t^\omega(\theta)\right)^4 \left(t^{2\omega}(\theta)\right)^2 \left(I^\omega\right)^2 \left[1 + \cos\psi\right] \tag{2}$$

$$\psi = \left[2\frac{\omega}{c}\left(L\left(n^{2\omega}\cos\theta^{2\omega} - n^\omega\cos\theta^\omega\right) + s\left(n_o^{2\omega}\cos\theta_o^{2\omega} - n_o^\omega\cos\theta_o^\omega\right)\right)\right]$$

where $t^\omega(\theta)$ and $t^{2\omega}(\theta)$ are transmission factors, n^ω and $n^{2\omega}$ are the refractive indices of the LB-film, n_o^ω and $n_o^{2\omega}$ are the refractive indices of the pyrex substrate, θ is the angle of incidence, θ^ω, $\theta^{2\omega}$, θ_o^ω and $\theta_o^{2\omega}$ are the internal angles in the LB-film and the substrate respectively, L and s are the thicknesses of the DCANP layer and the substrate. The phase factor ψ determines the form of the fringe pattern. Since $L \ll s$ the minima of the interference figures are mainly determined by the refractive indices of the substrate. The envelope of the curve depends on $t^\omega(\theta)$ and $t^{2\omega}(\theta)$ and therefore on the refractive indices of DCANP. Fig.3 shows the experimental points and a fit using equation (2). The results of such fits are summarized in the table.

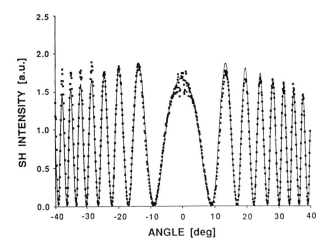

Fig 3.   Experimental curve of SHG as a function of the angle of incidence for a coating on both sides. The theoretical curve is overlayed.

The measurements were performed on different samples and different film thicknesses. The larger value for d_{333} can be explained by considering the molecular alignment. The charge-transfer axis of the chromophores which has the largest hyperpolarizability lies in a plane parallel to the dipping direction X_3.

4. ORIENTATION OF THE MOLECULES

When rotating the sample around X_1 and using a fundamental beam with the polarization direction perpendicular to X_1 we observed an asymmetry of the experimental curve of $I^{2\omega}(\theta)$ peaking at an angle of $\theta = (20 \pm 10)$ degrees (internal angle). This result indicates that most likely the charge-transfer axis of the chromophore is tilted by (20 ± 10) degrees with respect to the substrate surface. This means also that the alkyl-chain is tilted with respect to the chromophore. Further investigations concerning the orientation of the molecules in mono- and multilayers are in progress.

These measurements are in agreement with our birefringence experiments. There we obtained values of $\Delta n = n_1 - n_3 > 0$. If the charge transfer axis of the molecules were parallel to the substrate plane a value $\Delta n < 0$ would have been expected due to a larger polarizability of the molecule along the long molecule axis.

5. WAVEGUIDING EXPERIMENTS

One of the most interesting features of the LB technique is the ability to obtain thin films of a very accurately determinable thickness. This feature makes these films very attractive for guided-wave applications (Pitt *et al* 1976). In our experiments we used gratings (grating constant $\Lambda = 0.38$ μm) to couple into the DCANP waveguide. We preferred the use of grating couplers since mechanical problems may arise when clamping a prism strongly to

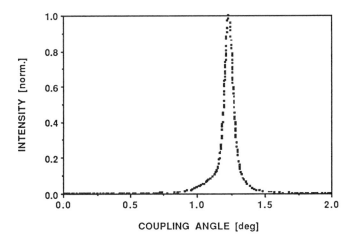

Fig. 4. Normalized coupling curve of a TE mode into a DCANP waveguide.

the LB film. The grating was first etched in the pyrex substrate by reactive ion etching. Subsequently the LB film was deposited with the dipping direction parallel to the grating grids. For a layer thickness of 220 nm a He-Ne laser beam ($\lambda = 633$ nm) with polarization parallel to the grating grids was coupled into the DCANP waveguide for guided wave TE mode propagation. The measured coupling angle $\alpha = -9.55°$ lead to an effective index $N_{eff} = 1.505$ where $n_s < N_{eff} < n_3$ (n_s = refractive index of substrate) Fig.4 shows the corresponding coupling curve. We observed propagation distances of up to 15 mm. Better gratings with a very low edge depth will be designed for further work in order to overcome the perturbation of the LB film structure by the grating. Dispersion, orientational effects between film and grating, attenuation of the guided wave intensity and frequency-doubling efficiencies are currently investigated.

Table : Optical properties of DCANP

Transparency range	400 nm - 2000 nm
Absorption maximum	$\lambda_{max} = 375$ nm
Absorption coefficient at λ_{max}	$\alpha_\| = 127\ 000$ cm^{-1}, $\alpha_\|/\alpha_\perp = 1.6$
Birefringence ($\lambda = 550$ nm)	$\Delta n = 0.056$
Nonlinear optical susceptibilities	$d_{333} = 6.8 \pm 1.2$ pm/V $d_{311} = 0.9 \pm 0.2$ pm/V
Effective refractive index for - $\lambda = 633$nm - thickness d = 220 nm - coupling angle $\alpha = -9.55°$ - grating constant $\Lambda = 0.38$ μm	$N_{eff} = 1.505$

6. CONCLUSIONS

The results described above demonstrate an enhanced second-harmonic generation efficiency for Y-type LB-films of a single material with a thickness up to 1.2 μm. The films are highly ordered. The order within the bilayer is most likely introduced by the dipping process and stabilized by interlayer hydrogen bridging between the amino and nitro functional groups of adjacent molecules. This type of hydrogen bridging is known to exist in single crystals of COANP (Bosshard *et al* 1989). The anisotropy of the LB-film is verified by the observed birefringence and by the strong polarization dependence of second-harmonic generation. The first waveguiding experiments with long propagation distances are promising. They indicate a good optical homogeneity of the DCANP films.

7. ACKNOWLEDGEMENTS

We thank G. Decher and S. Remund for the film preparation and the Swiss National Science Foundation (NFP 19: Materials for Future Technology) for financial support.

8. REFERENCES

Armstrong J A, Bloembergen N, Ducuing J, Pershan P S 1962 *Phys. Rev* **127** 1918
Bosshard Ch, Sutter K, Günter P and Chapuis G 1989 *JOSA B* **6** (4) 721
Decher G, Tieke B, Bosshard Ch and Günter P 1988 *J. Chem. Soc., Chem. Commun.* **1988**, 933
Günter P, Bosshard Ch, Sutter K, Arend H, Chapuis G, Twieg R J and Dobrowolski D 1987 *Appl. Phys. Lett.* **50** (9) 486
Kuhn H, Möbius D and Bücher H, 1972 *Physical Methods of Chemistry* Vol. 1, Part IIIB, Editors : Weissberger A and Rossiter P, Wiley Interscience p. 577
Pitt C W, Walpita L M 1976 *Electron. Lett.* **12** (18) 479
Shen Y R 1984 *The Principles of Nonlinear Optics* (John Wiley and Sons, New York) pp 98-101
Sutter K, Bosshard Ch, Ehrensperger M, Günter P and Twieg R J 1988 *IEEE J. Quantum Electron.* **24** (12) 2362
Zyss J 1985 *J. Mol. Electron.* **1** 25

A new non-linear optical Langmuir-Blodgett film material based on a liquid crystalline butadiyne

J Tsibouklis[ab], J Cresswell[a], C Pearson[a], M C Petty[a] and W J Feast[b]

[a]Molecular Electronics Research Group, School of Engineering and Applied Science, University of Durham, Science Laboratories, South Road, Durham DH1 3LE, U.K.

[b]Department of Chemistry, University of Durham, Science Laboratories, South Road, Durham DH1 3LE, U.K.

ABSTRACT: A novel, conjugated, unsymmetrically disubstituted, mesogenic diphenylbutadiyne capable of forming Langmuir-Blodgett films has been synthesised and characterised. Surface plasmon resonance experiments have indicated that monolayers of the material possess a second-order non-linear optical coefficient approaching that obtained from a well studied hemicyanine dye.

1. INTRODUCTION

Second order non-linear optical (NLO) effects are manifested by highly polarisable molecules capable of arranging themselves in a non-centrosymmetric manner. However these factors are not always mutually compatible; in most instances, dipole-dipole interactions between molecules, which individually possess large second-order hyperpolarisabilities, result in the formation of a more energetically favourable centrosymmetric crystal structure.

The Langmuir-Blodgett (LB) technique represents one of the most easily accessible methods for the engineering of non-centrosymmetric molecular structures. Monolayer and multilayer (superlattice) LB systems exhibiting significant second-order NLO effects may readily be fabricated (Allen et al 1987 and Neal et al 1986).

In previous work we have reported the synthesis and non-linear optical behaviour of a series of functionalised diarylalkynes which are capable of forming non-centrosymmetric LB films by Z-type deposition (Tsibouklis et al 1989). We now report some preliminary studies of the liquid crystalline behaviour, LB film formation and NLO properties of the diaryldialkyne (I) in which the chromophore has been extended by the introduction of a second triple bond.

$$NO_2-\langle O \rangle-C\equiv C-C\equiv C-\langle O \rangle-N=CH-\langle O \rangle-O-(CH_2)_{11}CH_3 \qquad (I)$$

In common with other substituted butadiynes (Wegner 1969) this compound lends itself to solid state polymerisation via a trans C_1-C_4 addition reaction as shown below:

$$nR_1-C\equiv C-C\equiv C-R_2 \longrightarrow \left\{\begin{matrix} R_1 \\ | \\ C-C\equiv C-C \\ | \\ R_2 \end{matrix}\right\}_n \quad \underline{OR} \quad \left\{\begin{matrix} R_1 \\ | \\ C=C=C=C \\ | \\ R_2 \end{matrix}\right\}_n$$

Such polymerisation reactions may be initiated in appropriate systems by exposure of the monomer crystals or LB films (Tieke 1985) to UV-light, γ-irradiation, certain reactive gases, heat, or elevated pressure.

2. EXPERIMENTAL

2.1 Synthesis of (I)

The diphenylbutadiyne (I) was synthesised from 4-nitro-4'-aminodiphenyl-butadiyne (Tsibouklis 1988a,b) according to the following method:

4-Nitro-4'-aminophenylbutadiyne (0.26 g, 0.01 mol) and 4-n-dodecyloxy-benzaldehyde (0.59 g, 0.02 mol) in methanol (20 cm^3) were stirred at 50°C until the colour of the mixture changed from orange to yellow (about 3 hrs). The mixture was filtered and washed with hot methanol (3 x 70 cm^3) to give (0.51 g, 0.0096 mol, 96%) of (I) as a yellow microcrystalline solid, m.p. 98°C (L.C.);   found C, 78.49;   H, 7.07;   N, 5.14%;   M$^+$ $_{mass}$ $_{spec.}$ 535 $C_{35}H_{38}N_2O_3$ requires C, 78.60 ; H, 7.17; N, 5.24%; M, 535; ν 2205 cm^{-1} (-C≡C-); λ_{max} 396 nm (tetrahydrofuran); ^{1}H-NMR (CDCl$_3$), [w.r.t. internal TMS] 8.3 (s,1H);   8.2 (d,2H);   7.8 (d,2H);   7.6 (d,2H); 7.5 (d,2H);   7.1 (d,2H);   6.9 (d,2H);   4.0 (t,2H);   1.8 (t,2H);   1.4 (t,2H);   1.2 (s,16H);   0.8 (t,3H). The purity of compound (I) was checked by HPLC on an ODS column with THF as solvent (room temperature; flow rate: 1 cm^{3}min^{-1}), a single narrow peak was displayed.

2.2 Film deposition and characterisation

Monolayer studies were conducted using a constant-perimeter trough which has been described previously (Petty 1987). The material was spread from solutions (ca. 1 mgcm^{-3}) in chloroform (BDH, Aristar grade) onto a water subphase at approximately 20°C. The water was purified by reverse osmosis followed by deionisation and UV sterilisation. During the measurement of isotherms and monolayer deposition, the water pH was 5.6-5.8.

Surface plasmon resonance (SPR) studies were performed with monolayers deposited onto silver-coated glass slides.

The NLO behaviour of the monolayer structures was explored using a ATR cell (Cross et al 1987). An applied ac field (3 kHz) modulated the cell reflectivity by varying the permittivity of the monolayer.

3. RESULTS AND DISCUSSION

3.1 Liquid-crystalline behaviour

In common with other butadiynes of this type (Tsibouklis 1988a) compound
(I) possesses liquid crystal phase transitions.   Above 220°C when the
material is in a nematic phase, an irreversible liquid crystalline state
polymerisation process takes place.   The DSC record obtained from (I) is
shown in figure 1.

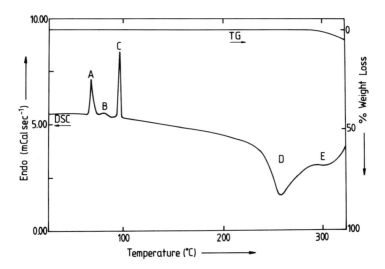

Figure 1.   DSC thermogram of the liquid crystalline butadiyne, (I);
6.7 mg;   first heating;   heating rate 5°C min^{-1}.   A,B: crystal-crystal
transitions;   C: crystal-smectic liquid crystal transition;   there is a
smectic to nematic transition at ca. 215°C which is masked by the onset of
polymerisation;   D: polymerisation;   E: degradation.   The TG trace is
also shown.

Evidence for the formation of a liquid crystalline monomer and polymer was
obtained by optical microscopy as shown below:

The polymer obtained from the liquid crystalline state polymerisation of (I) had a black metallic appearance and produced elemental analysis results consistent with a homopolymerisation of (I). In common with other diarylbutadiynes which polymerise in the solid state (Wegner 1971) the major difference between the infrared spectra of the monomer and the polymer is that the absorption of the triple bond at 2205 cm^{-1} disappears on polymerisation   giving place to a weak and broad absorption at 2180 cm^{-1}. ^{1}H-NMR, UV/Vis and mass spectroscopic investigations suggest that the polymer has a comb-like structure similar to that observed in other side chain, liquid crystalline polymers (Finkelmann 1983);   further details will be reported elsewhere.

3.2 Film formation

The pressure versus area curves, measured at compression rates of ~2 cm^{2}s^{-1} for compound (I) are presented in figure 2;   curve A represents the isotherm obtained approximately 5 minutes after monolayer spreading, while B is that obtained on subsequent recompression.   After recompression, the floating layer was found to be somewhat rigid.   In fact, the behaviour of the layer was reminiscent of that observed for many phthalocyanine films (Baker 1985).   The condensed area per molecule for curve B in figure 2 is approximately 0.20 nm^2.   This corresponds to that expected from the cross-section of the hydrocarbon chain.   Thus a high degree of packing for the chromophore units in the material is apparent.   It should be noted that isotherms similar to those in figure 2 were only obtained if the monolayer was spread from a <u>fresh</u> solution.   Pressure-area curves measured for solutions which were more than a few days old were much more expanded and invariably exhibited a plateau at about 20 mNm^{-1}.   This occurred despite the fact that the solution was always stored in the dark and in a refrigerator.   Further studies into this phenomenon, which may be linked to polymerisation, are in progress.

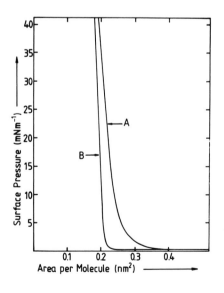

Figure 2.   Pressure versus area isotherm for (I).

Monolayers of (I) could readily be transferred to hydrophilic glass or to silver-coated glass slides at a deposition rate of 2 mmmin^{-1}. The transfer ratio was approximately 1. Multilayer deposition proved more difficult. For hydrophilic glass, good pick-up was achieved on the substrate upstroke, but the material was removed on the downstroke; in the case of hydrophobic glass (i.e. treated with a silanising agent), the opposite effects were noted. However, some success was found if the substrate was first coated with a few layers of a fatty acid. In view of these problems, the discussion of the non-linear optical properties will be restricted to monolayer samples.

3.3 NLO characterisation

Figure 3 shows reflectivity data obtained from a preliminary investigation into the NLO properties of a monolayer of (I) deposited onto a silver-coated glass slide. The dc curve is given in the inset; the differential signal was obtained using an applied voltage of 30 V rms across an air gap of 1.5 μm. The magnitude of the differential signal is approximately one third of that measured for a monolayer of a hemicyanine dye (Neal et al. 1986) measured under identical conditions and is significantly larger than that reported previously by us for the related diarylalkyne (Tsibouklis et al. 1989). This represents an exciting result when one considers that the material reported here (a) may be capable of polymerisation and (b) may also exhibit third-order NLO properties.

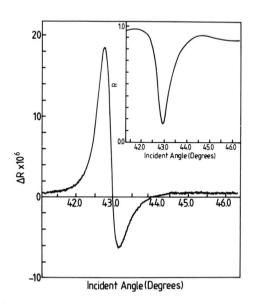

Figure 3. Differential reflectivity measured for one monolayer of (I). Inset shows d.c. SPR curve.

4. CONCLUSION

Compound (I) is a new NLO material which contains a hydrophobic moiety and a delocalised π-electron system substituted with donor and acceptor groups. The molecule is promising from the "molecular engineering" point of view since it can be incorporated into organised supermolecular arrangements such as liquid crystals and Langmuir-Blodgett films. More detailed studies on the liquid crystalline behaviour, LB film formation characteristics and NLO properties of (I) are now in progress.

ACKNOWLEDGEMENT

We thank the SERC for supporting this work.

REFERENCES

Allen S, Hann R A, Gupta S K, Gordon P F, Bothwell B D, Ledoux I, Vidakovic P, Zyss J, Robin P, Chastaing E and Dubois J C 1987 *Proc. Soc. Photo-Opt. Eng.*, **682** 97.
Baker S 1985 Ph.D. thesis, University of Durham.
Cross G H, Girling I R, Peterson I R, Cade N A and Earls J D 1987 *J. Opt. Soc. Am. B* **4** 962.
Finkelmann H 1983 '*Liquid Crystals*: *Their Physics, Chemistry and Applications*' ed C Hilsum and E P Raynes (Cambridge: University Press) pp 35-44.
Neal D B, Petty M C, Roberts G G, Ahmad M M, Feast W J, Girling I R, Cade N A, Kolinsky P V and Peterson I R 1986 *Electron. Lett.* **22** 460.
Petty M C 1987 '*Polymer Surfaces and Interfaces*' ed W J Feast and H S Munro (Wiley) 163.
Tieke B 1985 *Adv. Polym. Sci.* **7** 79.
Tsibouklis J, Werninck A R, Shand A J and Milburn G H W 1988a *Liquid Crystals* **3(10)** 1393
Tsibouklis J, Werninck A R, Shand A J and Milburn G H W 1988b *Chemtronics* **3** 211.
Tsibouklis J, Cresswell J, Kalita N, Pearson C, Maddaford P J, Ancelin H, Yarwood J, Goodwin M J, Carr N, Feast W J and Petty M C *J. Phys. D* in press.
Wegner G 1971 *J. Polym. Sci., Polym. Lett. Ed.* **9** 133.
Wegner G 1969 *Z. Naturforschg.* **24b** 824.

Inst. Phys. Conf. Ser. No 103: Section 2.4
Paper presented at Int. Conf. Materials for Non-linear and Electro-optics, Cambridge, 1989

Measurement of thermal and reorientational third order optical nonlinearities in liquid crystal fluids with absorption

K J McEwan and K J Harrison
Royal Signals and Radar Establishment
St Andrews Road
Great Malvern
UK

and
P A Madden
Physical Chemistry Laboratory
University of Oxford
South Parks Road
Oxford
UK

Abstract

The various contributions to $\chi^{(3)}$ in liquid crystals with absorption are analysed using D4WM on the nanosecond timescale. The contributions are examined by measuring the four nonzero elements of $\chi^{(3)}$. The experimental results are compared to the theoretical values and are found to agree well.

Introduction

The experimental technique of Degenerate Four Wave Mixing (D4WM) can be used to measure the third order nonlinear susceptibility ($\chi^{(3)}$) in liquid crystals, Madden et al (1986). In this paper we describe the measurements of $\chi^{(3)}$ in liquid crystals with absorption. Two laser beams are brought together simultaneously and coincidentally inside a material which responds nonlinearly, as shown in Figure 1. In response to the laser interference pattern the material forms a refractive index or phase grating which can be simultaneously "read out" by diffracting a third beam from the grating. The diffracted beam will contain information describing the amplitude of $\chi^{(3)}$ and its relaxation characteristics.

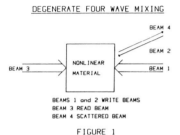

DEGENERATE FOUR WAVE MIXING

BEAM 4
BEAM 2
BEAM 3
NONLINEAR MATERIAL
BEAM 1

BEAMS 1 and 2 WRITE BEAMS
BEAM 3 READ BEAM
BEAM 4 SCATTERED BEAM

FIGURE 1

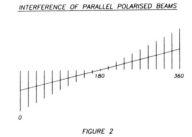

INTERFERENCE OF PARALLEL POLARISED BEAMS

0   180   360

FIGURE 2

If it is assumed that the two interfering beams have the same wavelength and are coherent, than at any position in the overlap region the beams will have a constant phase difference.   This results in a stationary interference pattern and therefore a stationary refractive index or phase grating.   With the amplitude of the two beams equal the interference at any position depends upon the phase difference and the polarisation of the beams.   If the polarisations are equal, the beams interfere constructively when the phase difference equals 0, 2π, 4π and destructively for π, 3π 5π as shown in Figure 2.   This type of grating is known as an "intensity grating". The situation is more complex where two orthogonally polarised beams interfere.   The intensity remains constant throughout the overlap region, but the local direction of oscillation of the electric field vector changes.   At positions where the phase difference equals 0, π, 2π the resultant electric field oscillates linearly at an angle subtending the two incident polarisations, and where the phase difference equals $\pi/2$, $3\pi/2$, $5\pi/2$ the resulting local polarisation will be circular.   This type of grating is known as a "reorientation grating" and is shown in Figure 3.

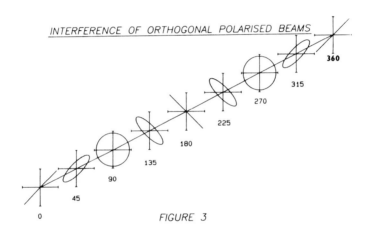

INTERFERENCE OF ORTHOGONAL POLARISED BEAMS

FIGURE 3

Theory

In this study pulses of approximately 25 ns FWHM are used to examine organic liquids of anisotropic molecules, some of which are nematogenic.   In this time regime there are two dominant mechanisms which contribute to $\chi^{(3)}$.

1)   Molecular reorientation, which occurs where anisotropically polarisable molecules reorient under the influence of an optical field to align their most polarisable axis with the polarisation of the electric field..

2)   The thermal effect, which relies on density changes resulting from temperature fluctuations caused by molecular absorption of the radiation, Hoffman (1986).

Since the contribution to $\chi^{(3)}$ from each mechanism depends on the type of grating, it is possible to measure the contribution to $\chi^{(3)}$ from the two different nonlinear mechansims by examining various · polarisation combinations.   The polarisation combinations examined and the theoretically expected contribution to $\chi^{(3)}$ from reorientation and thermal mechanisms are shown in Figure 4.

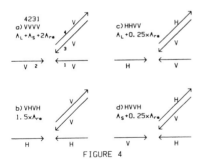

FIGURE 4

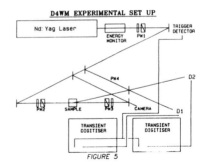

FIGURE 5

Experimental

The experimental set up is shown in Figure 5. The Nd:Yag laser source (1.06 μm) delivered a Q–switched pulse of FWHM 25 ns, in a single longitudinal mode and approximate TEM_{00} spatial mode of $1/e^2$ diameter about 3 mm. An energy level of 100 mJ per pulse was normally employed. However control of the total energy, and the polarisation and energy of each beam was provided by calibrated polariser waveplate pairs (PW). The nonlinear samples were contained in a 1 cm path length cell.

All the information required to describe the nonlinear process is contained in the temporal profile of the diffracted pulse. By measuring both the diffracted temporal profile and the input temporal profiles using detectors D1 and D2, then using a computer fitting routine, the parameters describing the nonlinear behaviour can be determined McEwan et al (1989). Figure 6 shows a computer fit to the experimental diffracted pulse where only reorientation contributes. The reorientation time and the nonlinear amplitude corresponding to molecular reorientation are determined. Figure 7 shows the situation where both molecular reorientation and the thermal mechanisms contribute to $\chi^{(3)}$. The diffracted signal goes to zero because both mechanisms act in opposition. By using the reorientation time determined above, both nonlinear amplitudes and the thermal diffusion coefficient can be determined.

Results

The material chosen for experimental study were:

1)   Pentyl cyanobiphenyl (5CB) which is known to have a large reorientation contribution close to T_{N-I} Madden et al (1986).

2)   Chloroform which has a significant thermal signal, assigned to absorption of 1.06 μm radiation by an overtone of the C–H stretching vibration.

3)   20% Chloroform and 80% 5CB mixture which has comparable reorientation and thermal contributions.

The experimentally determined nonlinear amplitudes are compared to the theoretical values in Tables I, II and III.

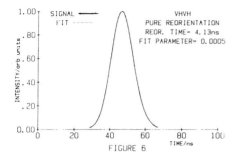

FIGURE 6

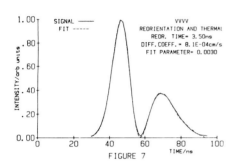

FIGURE 7

RESULTS

CYANOPENTYLBIPHENYL (5CB)

POLARISATION COMBINATION	MEASURED REORIENTATION AMLITUDE	THEORETICAL REORIENTATION AMPLITUDE
VVVV	1	1
VHVH	0.70±0.03	0.75

TABLE I

PURE CHLOROFORM

POLARISATION COMBINATION	MEASURED THERMAL AMPLITUDE	THEORETICAL THERMAL AMPLITUDE
VVVV	1	1
VHHV	0.97±0.04	1
HHVV	1.04±0.04	1

TABLE II

80% 5CB + 20% CHLOROFORM

POLARISATION COMBINATION	MEASURED REORIENTATION AMPLITUDE	THEORETICAL REORIENTATION AMPLITUDE	MEASURED THERMAL AMPLITUDE	THEORETICAL THERMAL AMPLITUDE
VVVV	1	1	1	1
VHVH	0.63±0.04	0.75	-	-
VHHV	0.10±0.024	0.125	1.08±0.07	1
HHVV	0.16±0.01	0.125	1.16±0.06	1

TABLE III

Conclusion

The agreement between the experimental results and the theoretically expected values, verifies the experimental technique and computer fitting routines. It has been shown that the technique can be used to measure both reorientation and thermal contributions to $\chi^{(3)}$ in liquid crystal materials with absorption.

References

[1]   Madden, P A, Saunders, F C, Scott, A M, Optical. Acta., 1986, vol 33, pp 405–417.

[2]   McEwan, K J, Harrison, K J, Madden, P A, to be published.

[3]   Hoffman, H J, Jour. Opt. Soc. Am. B, 1986, vol 3.

Inst. Phys. Conf. Ser. No 103: Section 2.4
Paper presented at Int. Conf. Materials for Non-linear and Electro-optics, Cambridge, 1989

Advances in fast electro-optical phenomena in liquid crystals

Kent Skarp

Physics Department, Chalmers University of Technology, S-412 96 Göteborg Sweden

ABSTRACT: Recent developments in fast electro-optical processes in liquid crystals will be discussed. Emphasis is put on chiral liquid crystal systems, in particular the ferroelectric tilted smectic phases. The electro-optical switching is analyzed in terms of current phenomenological models, and the influence of the rotational viscosity and other parameters on response times is considered. The electroclinic effect in orthogonal smectic phases is described, and results on the electro-optic switching characteristics of some novel materials will be presented.

1. INTRODUCTION

Liquid crystals have proven to be an increasingly important member of the expanding family of electro-optic materials over the last two decades. Until recently, however, the use of liquid crystals in devices was restricted to areas not requiring switching times much below 10-100 ms. Devices employing the familiar twisted nematic and supertwisted birefringence effects, and various developments of these, are today commonly found in many types of display. The main features of all these effects are that they use electric-field induced reorientations in a nematic liquid crystal, and that the coupling between the electric field and the director (the average molecular orientation) relies on the dielectric anisotropy, and thus, since it depends on the <u>induced</u> polarization, is a second-order effect in the field strength E.

In recent years, interest has focussed on phenomena that have their origin in the interaction between the externally applied electric field and <u>permanent</u> molecular dipoles. This area of liquid crystal reasearch was opened up by Meyer *et al* (1975), who predicted and demonstrated the existence of ferroelectricity in the chiral substance DOBAMBC, revealing the macroscopic dipolar order in this substance in its chiral smectic C phase (the C* phase). Today it is clear that there are at least seven tilted smectic phases that have the required symmetry for ferroelectric order when they exhibit chirality. Because there is a dipolar coupling to the applied field, the ferroelectric phases exhibit linear effects in the field strength E. Response times are short, in the 10-100 µs range for present materials. Several reviews of the field have been presented during the last years (Blinov and Beresnev 1984, Lagerwall and Dahl 1984, Skarp and Handschy 1988, Beresnev *et al* 1988).

More recently, also orthogonal chiral smectic phases have attracted attention due to the electroclinic effect, first described by Garoff and Meyer (1977). This is a linear effect and relies on the coupling of dipol moments to the applied field. One might regard the orthogonal phases as paraelectric, with a diverging susceptibility at the transition to the adjacent ferroelectric, tilted phase. Associated with this dielectric softening at the transition is an induced molecular tilt, which arises due to the symmetry breaking action of the applied field. The electroclinic effect, which in the smectic A* phase is the excitation of the soft mode, exhibits the fastest response times so far observed in a useful electro-optic effect in liquid crystals, presently well below one microsecond (Andersson *et al* 1987).

2. STRUCTURE AND SYMMETRY IN LIQUID CRYSTAL PHASES.

In the nematic phase, the rod-like molecules tend to align parallel, and a thermodynamical average of the distribution function for the molecular orientations yields a quantitative measure of the order. The symmetry of this nematic structure is $D_{\infty h}$, meaning that there is complete rotational degeneracy around one axis (the director), the orientation of which is given by external conditions (boundaries, applied fields, flow gradients etc.). The smectic A phase has the same point group $D_{\infty h}$, but differs from the nematic by being a layered structure: The molecules are on average normal to the layers, so that in this direction the medium has a solid-like elasticity, while the molecules flow freely parallel to the layers. Typical textures of these phases are shown in Figure 1.

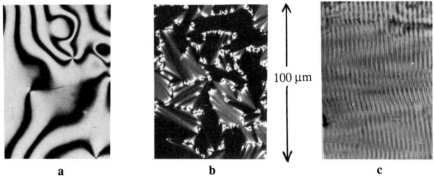

100 μm

a   b   c

Fig. 1. **a**. Texture of the nematic phase. **b**. Focal-conic texture of the smectic A phase. **c** Fingerprint texture of the C* phase resulting from the helicoidal structure.

Out of the existing 12 smectic classes, seven classes (C,I,F,G,H,J, and K) are characterized by a temperature-dependent tilt $\theta(T)$ between the director and the normal to the smectic layers, and they belong to point group C_{2h}. If additionally the constituent molecules are chiral (thus exhibiting optical activity), these smectic phases are denoted C*,I*,F*,G*,H*,J*, and K*, and all belong to point group C_2 allowing for a spontaneous polarization in the plane of the layers. Due to the molecular chirality, C*, F* and I* normally form a helicoidal structure with a pitch of a few microns (Figure 1c) that is conveniently studied by laser diffraction since it forms an optical phase grating.

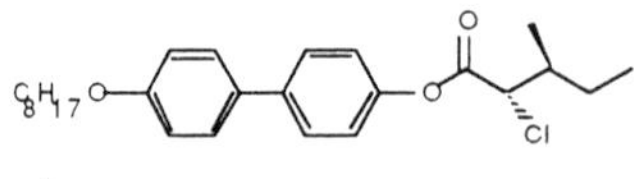

$C_8H_{17}O$ X 30 C* 53 A 66 L

Fig. 2. A chiral molecule forming a ferroelectric C* phase.

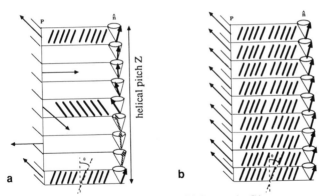

Fig. 3. Helicoidal and non-helicoidal smectic C* structures.

Because of the helicoidal structure of the C* phase, the spontaneous polarization averages out to zero macroscopically, cf. Figure 3a. By applying an electric field along the layers, or taking advantage of the role of surface conditions in thin cells , a polarized state can be obtained, as described by Clark and Lagerwall (1980), cf.Figure 3b. For applications, the smectic layers are arranged in bookshelf geometry between two glass plates coated with transparent ITO-layer, Figure 4. In a thin preparation (of thickness less than the helical pitch), the polarized states takes the form of domains, which are easily seen in a polarizing microscope, since the optic axis, lying in the plane of the cell, makes an angle of 2θ between the states. Because the birefringence is typically 0.1-0.2 and the tilt angle 15-25 degrees, one can obtain a contrast from extinction to first order white for a thickness of the liquid crystal layer of around 1-3 μm. Since the domains are with polarization UP or DOWN, it is possible to change their state by applying a field of proper polarity, cf Figure 4. For certain alignment techniques, it is furthermore possible to obtain a bistable operation between the two optical states.

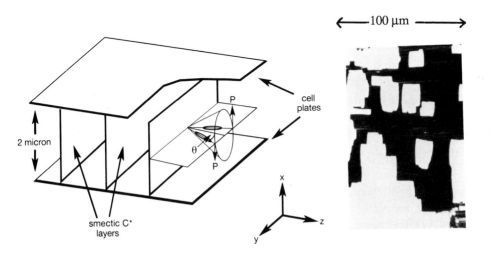

Fig. 4a Ferroelectric liquid crystal in bookshelf geometry with cone positions corresponding to the UP and DOWN domains in the micrograph of figure 4b..

3 EIGENMODES AND FAST RESPONSES IN THE C* AND A* PHASES.

The possibility to rotate the director on the smectic C cone (the Goldstone mode) by applying an external field is the basis for the application of C* phases for fast electro-optic applications. As shown in Figure 5, a short pulse of 10 μs duration will induce switching between the two extreme states on the cone, and the optical response (upper curve) shows that the effect exhibits bistable behaviour, with stable states between the electric pulses.

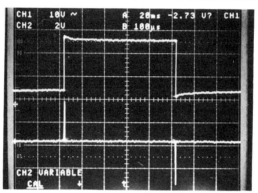

Fig. 5 Bistable electro-optic switching in a 2μm thick C* cell.

In a thick preparation, the helical structure is fully developed, and the structure has an eigenmode, called the Goldstone mode, for long-wave-length excitations. Excitations of this mode relax with a characteristic relaxation time τ_G determined by an elastic constant K_ϕ and a rotational viscosity γ_ϕ. The dynamical equation for the helix excitation is

$$K_\phi \sin^2\theta \frac{\partial^2\phi}{\partial z^2} - \gamma_\phi \sin^2\theta \frac{\partial\phi}{\partial t} = PE\sin\phi \qquad \tau_G = \frac{\gamma_\phi}{K_\phi q^2}$$

where ϕ is the azimuthal angle for the polarization P along z, and θ is the tilt angle (the angle between the director and the smectic layer normal). The equation has a characteristic time τ_G, usually of the order of 10-100 Hz, depending on the magnitude of the helical wave-vector $q=2\pi/Z$, where Z is the helical pitch.

In a thin preparation, the helix can be unwound by surface actions, and then the switching equation becomes

$$\gamma_\phi \sin^2\theta \frac{\partial\phi}{\partial t} = PE\sin\phi \qquad \tau_S = \frac{\gamma_\phi \sin^2\theta}{PE}$$

with the switching time τ_S determined by viscosity γ_ϕ, spontaneous polarization P and applied field strength E. Typical values in a modern FLC mixture is 20 nC/cm^2 for P and 0.2 Poise for the viscosity, which gives a switching time of 10 microseconds for an applied field of 10 V/μm. In a thin cell (usually of 1-2 μm thickness) with bistable operation, the switching occurs above a certain threshold given by pulse amplitude and width, i.e. the threshold is dynamic in contrast to the static (only field-dependent) threshold in nematics.

The molecular tilt angle θ is in the C* phase usually taken to be constant, only exhibiting a weak temperature dependence. Near the phase transition to the orthogonal A* phase, the tilt softens and becomes field-dependent. This soft-mode behaviour is especially pronounced in the paraelectric A* phase, and is the basis for the newest fast electro-optic effect in liquid crystals, which is called the electro-clinic or soft-mode ferroelectric effect (Andersson *et al* 1987). The induced tilt angle in the A* phase shows a linear field-dependence, meaning that the optic axis can be rotated in the plane of the sample by applying a field along the smectic layers, cf Figure 6.

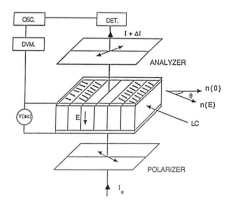

Fig.6 Soft-mode ferroelectric liquid crystal cell and set-up for determining its electro-optic characteristics.

The relation betwen the applied field and the induced tilt angle can be deduced from a Landau expansion of the free energy at the C*-A* transition (Andersson *et al* 1988), resulting in a linear relation

$$\theta = e_c E \qquad \text{where} \qquad e_c = \frac{\partial \theta}{\partial E} = \frac{\mu}{\alpha(T-T_c)}$$

where T_c is the A*-C* transition temperature and α and μ are Landau coefficients. The linear relation holds for small induced tilts, as seen from Figure 7a, but at large applied fields, or near the phase transition, a saturation in the tilt-field curve is observed, corresponding in principle to the dipolar saturation having the form of a Langevin function for a dipolar gas.

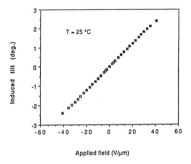

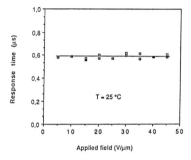

Fig. 7a Induced molecular tilt θ versus field due to the soft-mode effect

Fig. 7b. Response time versus field for the mixture Merck 88-158.

The response time of the soft-mode effect shows a field-independent behaviour, as seen in Figure 7b. From the Landau expression the response time in the A phase is given by

$$\tau_A = \frac{\gamma_\theta}{\alpha(T-T_c)}$$

where γ_θ is the rotational viscosity for the soft-mode (change of tilt angle). The response time is usually below one microsecond (cf. Figure 7b), but shows a critical behaviour near the A*-C* transition due to the critical slowing down. The effect can be considerably faster at temperatures away from the transition, and the measured response times will often be limited by the RC-time constant of the cell, determined in practice by the resistivity of the ITO layer. In non-transparent cells with metallic electrodes, response times down to 50 ns have recently been deduced from the dielectic soft-mode spectra.

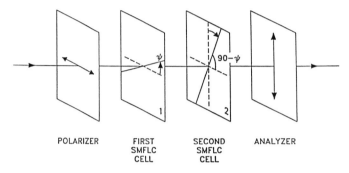

| POLARIZER | FIRST SMFLC CELL | SECOND SMFLC CELL | ANALYZER |

Fig. 8 An achromatic modulator with two soft-mode cells.

Due to the limited angular range available in present materials (an induced tilt of 10-12 degrees is the largest reported so far), it is necessary to use two cells in series to obtain a total rotation of 90 degrees for the polarization plane of incoming plane-polarized light, as in figure 8. The soft-mode cells are assumed to have a thickness and birefringence that make them act as $\lambda/2$ plates, with a rotatable optic axis. The soft-mode effect has a broad applicability for a number of different areas, including colour switches and fast modulators with full gray-scale capacity (Andersson *et al* 1989). The present practical drawbacks of a temperature-dependent response time and tilt angle are likely to be overcome by intensified efforts in organic chemistry, in a way that has become common in liquid crystal science from the nematics to the ferroelectric smectic C* materials that are now available in broad-range mixtures with well-specified physical parameters.

REFERENCES
Andersson G, Dahl I, Keller P, Kuczynski W, Lagerwall S T, Skarp K and Stebler B
     1987 *Appl.Phys.Lett.* **51** 640
Andersson G, Dahl I, Kuczynski W, Lagerwall S T, Skarp K and Stebler B 1988 *Ferroelectrics* **84** 285
Andersson G, Dahl I, Komitov L, Lagerwall S T, Skarp K and Stebler B 1989 *J.Appl.Phys.* (in press)
Beresnev L A, Blinov L M, Osipov M A and Pikin S A 1988 *Mol.Cryst.Liq.Cryst.* **158A** 1
Blinov L M and Beresnev L A 1984 *Sov.Phys.Usp.* **27** 492
Clark N A and Lagerwall S T 1980 *Appl.Phys.Lett.* **36** 899
Garoff S and Meyer R B 1977 *Phys.Rev.Lett.* **38** 848
Lagerwall S T and Dahl I 1984 *Mol.Cryst.Liq.Cryst.* **114** 151
Meyer R B, Liebert L, Strzelecki L and Keller P 1975 *J.Physique, Lett.* **36** L-69
Skarp K and Handschy M A 1988 *Mol.Cryst.Liq.Cryst.* **165** 439

Inst. Phys. Conf. Ser. No 103: Section 2.4
Paper presented at Int. Conf. Materials for Non-linear and Electro-optics, Cambridge, 1989

203

Electro-optic and nonlinear optical devices using liquid crystals

J Staromlynska, D W Craig, R A Clay, A Miller, K J Harrison & D Tytler*

Royal Signals & Radar Establishment, Gt Malvern, Worcs WR14 3PS

ABSTRACT: This work describes studies carried out across the visible on electro–optic and nonlinear optical devices incorporating a range of liquid crystal materials. Results are presented on electro–optically tunable filters in Fabry Perot and variable birefringence filter (VBF) geometries. Optical bistability in a Fabry Perot etalon with metal mirrors is demonstrated and a novel device which brings together the EO and nonlinear effects to produce bistability of the optical output with applied voltage is described.

1. INTRODUCTION

Liquid crystals are a class of materials which exhibit very large electro–optic and nonlinear coefficients (Raynes 1987 and Staromlynska et al 1987). It is easily possible to tailor the magnitude of both of these effects by changing the molecular structure of the liquid crystal mixture, (Bradshaw et al 1983). In general most liquid crystals may be used over a very large spectral range. In addition LC device technology is mature and inexpensive and is readily adaptable to the production of optical components such as tunable VBFs or tunable Fabry Perot resonant cavities. Liquid crystals thus offer much potential for the control and modulation of laser light through the electro–optic or nonlinear effects or both.

This work describes studies carried out across the visible on state–of–the–art devices incorporating a range of LC materials (K15, E44, 16522 and 3/5/7 PCH). Results are presented on electro–optically tunable VBFs containing each of these four materials. Extinction ratios as large as 1000: 1 are reported and it is shown that the larger the material birefringence the larger is the tuning range. A direct comparison of a VBF is made with a FP etalon with dielectric mirrors, both containing the material K15. The merits of each geometry are discussed and the power handling capability of each device is included. A study was made of the influence of alignment on tuning range and response times. It is shown that pre–tilting the liquid crystal reduces the tuning range and increases the response times.

Finally optical bistability in a Fabry Perot etalon with metal mirrors is demonstrated. A series of bistable loops obtained at different initial cavity detunings obtained by electrically fine tuning the etalon are shown. A novel device which brings together the electro–optic and nonlinear effects to produce bistability of the optical output with applied voltage is described. Typical characteristics of this device are presented and it is shown that if dn/dv and dn/dI are of the same sign then the 'voltage bistable' loops follow an anti–clockwise path.

* Present Address: School of Physics, Bath University, Bath, BA2 7AY

2.   EXPERIMENTAL DETAILS

Figure 1 shows the basic geometry of both the VBF and FP etalon devices.   In the VBF geometry the mirrors are absent.   All cells were constructed from ITO coated glass substrates.   The ITO acted as a rugged transparent electrode.   Production of the FP etalon involved molecular beam deposition of high–low pairs of BaF_2/ZnSe, λ/4 reflector stacks.   The metal mirrors were evaporated silver films of approximately 200A thickness. To form the VBF the alignment layer was deposited directly onto the ITO glass.   Two standard alignment agents were used in this work, SiO and polyimide.   The SiO yields an average tilt angle of the long axis of the molecules to the plane of the substrates of 12–15 degrees the polyimide yields approximately 2 degrees (Raynes 1987).

All measurements were made CW using a Krypton ion laser.   Figure 2 is a diagram of the experimental geometries adopted for the EO and bistability measurements.   Large area silicon photodiodes (D1 and D2) monitored the optical powers at the appropriate positions in the experimental arrangements.   Electro–optic modulation of the optical signal using the VBF was achieved by placing the cell at 45 degrees between crossed polarisers and using it as a tunable λ/2 plate.   Modulation with the FP etalon was achieved by aligning the molecules parallel with the polarisation of the input beam and monitoring the optical output immediately after the LC cell.   The results were taken by ramping the bias voltage across the sample and recording the transmission versus voltage characteristics on an X–Y recorder.   All voltages were AC at a frequency of 10 KHz.   The beam diameter was 1 mm.   The optical bistability loops were obtained by focussing the beam to a spot size of 12 microns ($1/e^2$ radius) and ramping the optical input using an AO modulator.   The 'voltage bistable' loops were recorded by fixing the optical input power and recording the optical output as a function of bias voltage.

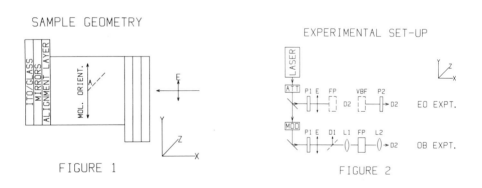

SAMPLE GEOMETRY   EXPERIMENTAL SET-UP

FIGURE 1   FIGURE 2

3.   RESULTS OF THE ELECTRO–OPTIC STUDIES

Modulation of light on the application of a bias voltage across the cell occurs through a change in the refractive index, (or birefringence), brought about by the re–orientation of the LC molecules.   All of the LC materials used in this work were of the nematic type with a positive dielectric anistropy which means that the molecules align with their long axis parallel to the direction of an applied electric field.   For the case of a VBF at 45 degrees between crossed polarisers, the transmitted intensity I_T is described by the equation

$$I_T = I_o \sin^2 \delta/2 \tag{1}$$

where I_0 is the incident intensity and δ is the phase retardance induced by the birefringent material. The number of 2π phase changes, δm, which take place on a change in birefringence, Δb, is given by

$$\Delta m = d \, \Delta b / \lambda \qquad (2)$$

where d is the device thickness and λ is the wavelength of the incident radiation. For a LC material the maximum possible change in birefringence is almost equal to the birefringence itself and hence equation 2 reduces to

$$\Delta m \text{ (TOTAL)} \cong db/\lambda \qquad (3)$$

Full reorientation of the molecules does not occur and hence Δb is always slightly less than b. Figure 3 shows the transmission versus voltage characteristics of the system for the four materials at $\lambda = 647$ nm. Figure 4 is the companion to Figure 3 and shows the response of E44 at two different wavelengths. These characteristics reflect the nonlinear change in birefringence with applied voltage. This is due to the nonlinear change of n in the y–direction with angle A, (Figure 1), plus the nonlinear change in A with applied voltage. The change in n with A is given by

$$n(A) = \frac{n_o \, n_e}{(n_e^2 \sin^2 A + n_o^2 \cos^2 A)^{1/2}} \qquad (4)$$

where n(A) is the effective refractive index in the y–direction, A is the tilt angle and no and ne are the ordinary and extraordinary refractive indices of the liquid crystal. At low volts the molecules in the centre of the cell reorient, at higher voltages the molecules close to the cells walls re–align. The change in n with applied voltage is discussed in the paper by Raynes 1987. Table 1 compares the Δm values predicted by equation 3 with the experimental values taken from the characteristics and calculated using equation 1. The errors quoted on the predicted values arise from the accuracy of the estimated cell thickness used in the calculation. In all cases good agreement was obtained between the predicted and experimental values.

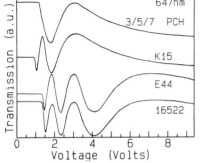

FIGURE 3: Transmission spectra for different materials

FIGURE 4: Transmission spectra for different λ

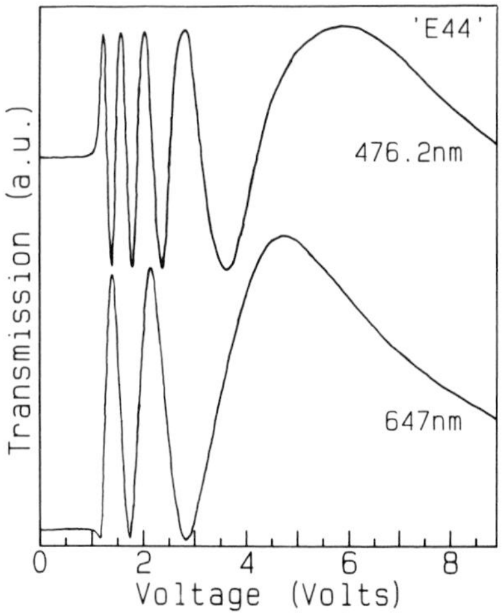

MATERIAL	b	LAMBDA	Δm pred.	Δm expt.
3/5/7/PCH	0.12	647nm	1.4±.1	1.4
K15	0.18	'	2.1±.2	2.1
16522	0.29	'	3.3±.3	3.1
E44	0.26	'	3.0±.3	2.9
E44	'	531nm	3.7±.4	3.8
E44	'	476nm	4.1±.4	4.5

TABLE 1. PREDICTED @ EXPT. VALUES OF Δm

An important parameter for a tunable filter is the dynamic range or contrast ratio. Contrast ratios using the VBF geometry with the four different materials were typically of the order 600-800:1. Typical maximum percentage transmission values were 70-80%. It was possible to obtain ratios of greater than 1000:1 by careful choice of sample or sample area. Poorly aligned samples (or areas of sample) resulted in the contrast ratio dropping to 200-300:1. The contrast ratio was not found to be wavelength dependent. The polarisers were cube beam splitters and when crossed yielded an extinction of 3.5×10^4

The performance of a FP tunable filter with dielectric mirrors was studied and compared to that of a VBF with the same material (K15). Figure 5 shows the response of such a filter at 647 nm. The transmission of a FP etalon is related to the double pass phase shift, δ, via the Airy function

$$T = A/(1 + F \sin^2 \delta/2) \qquad (5)$$

where A is a constant which depends on the mirror reflectivities, absorption and etalon thickness and F is the coefficient of finesse. The number of 2π phase changes on a change in refractive index of Δn follows the relation

$$\Delta m = 2\Delta n \ d/\lambda \qquad (6)$$

The total phase change hence approximates to

$$\Delta m \ (TOTAL) \cong 2bd/\lambda \qquad (7)$$

The greater number of fringes for the etalon reflects the greater sample thickness, (10 microns), and the factor of 2 difference between equations 7 and 3. Again good agreement was obtained between the predicted and experimental values of Δm. The best contrast ratio achieved with the etalon was 7:1 – ie well below the worst value obtained with the VBF. This highlights the fact that the etalon is a multiple beam interference geometry whereas the VBF is a single pass device. Any variation in cavity length, alignment etc is accentuated in the FP geometry and consequently the specifications are more stringent for the etalon than the VBF.

The power handling capability of each geometry was tested by recording the transmission versus voltage characteristics at a variety of optical input powers. Figures 5 and 6 show the results taken at λ = 647 nm. The characteristics of the etalon are strongly affected by the optical input power with the fringe pattern shifting to lower voltages with increasing optical power. This was first discernible at ~50 mW. A shift of ~ 2/3 of a fringe (corresponding to a Δn of 0.065) occurs at 300 mW. There is no obvious shift in the fringe pattern of the VBF until the optical input power is 300 mW. For λ = 531 nm the nonlinearity cuts in at ~ 40 mW for the etalon but remains unchanged for the VBF. The shifting of the fringes is due to the optical thermal nonlinearity of the LC, the change in n being brought about by the thermal disruption of the molecular ordering. dn/dI and dn/dV are of the same sign and hence the fringe pattern shifts to lower voltages. Heating occurs through absorption of the incident beam. The wavelength dependence of the nonlinearity in the FP geometry reflects the absorption profile of the ZnSe present in the mirrors (Smith et al 1984).

FIGURE 5: Power dependence of
the F-P tunable filter

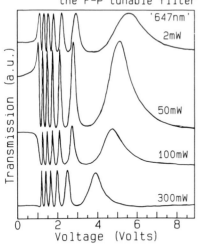

FIGURE 6: Power dependence of the
var. birefringent filter

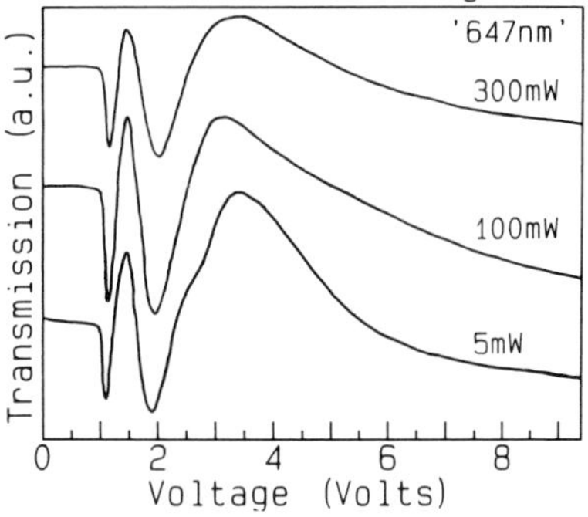

FIGURE 7: Effect of Pre-tilt

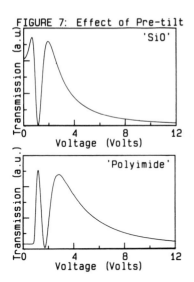

Figure 7 compares the responses of two VBFs one of which has an initial tilt angle of ~ 2° induced by polyimide and one which has a tilt of ~ 15° induced by SiO. There are two main features of the 15° pre-tilted device. The first is the much reduced threshold voltage before reorientation of the molecules (and hence modulation of the transmission) occurs and the second is the reduced tuning range. The reduced threshold voltage is due to the reduced restoring force of the alignment layer. The smaller tuning range is a consequence of the smaller total change in birefringence brought about by the 15° pre-tilt. The tuning range was calculated using equations 3 and 4 assuming an average tilt of 15°. This predicted $\Delta m = 3.36$ which agreed well with the experimental value of 3.2.

Measurement of the transmission maximum switching times revealed that the SiO aligned VBF had slower responses. A possible reason for this is that the SiO device was operating closer to the Freedricks transition (Raynes 1987) than the polyimide aligned cell. The response time of the LC reorientation follows the relation

$$\tau \propto \frac{1}{V^2 - Vc^2} \tag{8}$$

were τ is the response time, Vc is the threshold voltage and V is the applied voltage. Measurement of the approximate voltages for each device revealed $V^2 - Vc^2$ for the SiO aligned device was smaller than for the polyimide device.

4.   RESULTS OF THE BISTABILITY STUDIES

All of the bistability measurements were performed on the FP etalon with the partially reflecting metal mirrors containing the LC mixture 3/5/7 PCH. Two types of experiment were carried out. Firstly standard optical bistability loops were obtained by ramping the optical input power to the etalon. Figure 8 is a series of optical bistability loops obtained at different initial cavity detunings which were imposed by electrically fine tuning the cavity. Increasing the bias voltage tuned the cavity closer to a transmission maximum, this being reflected in the loops moving to a lower power and becoming narrower. In the second experiment the optical input power to the sample was held constant and the bias voltage was ramped. Figure 9 is a typical characteristic of the transmitted optical power versus bias voltage. This 'voltage bistable' loop corresponds to the cavity being tuned to a point where for the given input power switch–up (or down) can occur through the nonlinearity of the material. The loop follows an anti–clockwise path and is consistent with dn/dV and dn/dI being of the same sign. (Staromlynska et al 1987).

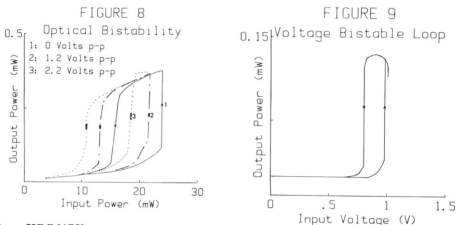

5.   SUMMARY

This work has helped to demonstrate the potential of LC materials for the control of laser light. All of the reported work was carried out using available LC materials and device technology. Further effort on both these fronts should result in an improved performance of the devices described.

ACKNOWLEDGEMENTS:–   We thank E P Raynes for useful discussions and EEV, Chelmsford for the supply of some of the LC cells.

REFERENCES

1.   Bradshaw M J, Constant J, McDonnel D G and Raynes E P, Mol Cryst Liq Cryst. 97, 177 (1983).

2.   Smith S D, Mathew J G H, Walker A C, Wherrett B S and Hendry A, Optics Comms, 51, 357, 1984.

3.   Raynes E P, "Electro–Optics and Photorefractive Materials" ed. P Gunter, Springer Proceedings in Physics Vol 18 1987.

4.   Staromlynska J, Miller A, and Clay R A, Optics Comms, 61, 415, 1987.

Inst. Phys. Conf. Ser. No 103: Section 2.4
Paper presented at Int. Conf. Materials for Non-linear and Electro-optics, Cambridge, 1989

209

Towards ordered conjugated polymers for non-linear optics: acetylenic and diacetylenic liquid-crystalline derivatives

J. Le Moigne, B. François, D. Guillon, A. Hilberer, A. Skoulios, A. Soldera and [*]F. Kajzar

Groupe des Matériaux Organiques,
Institut de Physique et Chimie des Matériaux de Strasbourg,
Institut C. Sadron, 6 Rue Boussingault,67083 STRASBOURG Cedex, France.
and
[*]DEIN/LPEM CEA Saclay
91191 Gif / Yvette Cedex, France

ABSTRACT:
Acetylenic and diacetylenic mesomorphic monomers have been synthesized in order to obtain long π-conjugated polymer chains able to give high third order optical non-linearities. First, the synthesis and liquid crystalline properties of several new diacetylenic monomers containing cholesteryl and methoxybiphenyl substituents are discussed. Second, the polymerization properties of acetylenic derivatives with cholesteryl pendant groups are presented. The conjugated polyacetylenic chains polymerized in the presence of transition metal halides have been characterized by standard polymer methodology (GPC, light scattering...), then studied from the standpoint of their optical properties.

1. INTRODUCTION

In recent years, an important effort has been devoted to the synthesis of novel monomers to obtain polymers with a large conjugated electron delocalization in the main chain. π-electron delocalization in a long conjugated polymer chain gives rise to unusual non-linear optical properties under high light intensity. Large third-order non-linear responses have been obtained with unsaturated polymers such as polyacetylenes (Kajzar et al 1987 a and b) or polydiacetylenes (Sauteret et al 1976, Carter et al 1985, Nunzi et al 1987).

High third-order susceptibility (χ^3) values in polymeric materials are strongly dependent on two main factors: the molecular hyperpolarizability and the degree of macroscopic orientation. Thus, single crystals of conjugated-chain polymers have been produced with symmetrical and unsymmetrical diacetylenic molecules. Macroscopic orientation can be also obtained through the alignment of molecules in a liquid crystalline state. In order to associate the properties of the conjugated backbone to those of polymers in the mesomorphic state, and more specifically of side-chain liquid crystalline polymers, several liquid crystalline diacetylenic monomers have been synthetized and polymerized (Garito et al 1984, Ozcayir et al 1984, 1986 (a) and (b), Schen 1988, Millburn 1988, Tsibouklis et al 1988).

In this paper, the synthesis and polymerization of new diacetylenic and acetylenic monomers incorporating cholesteryl and methoxybiphenyl mesogenic groups are reported and their structural and thermal behavior are briefly described. Some preliminary experimental results, including third-order non-linear susceptibilities, are given for a cholesteryl side-chain polyacetylene.

2. RESULTS AND DISCUSSION

A series of diacetylenic monomers is presented in table I:

Structure	K transitions
$CH_3-(CH_2)_7-C{\equiv}C-C{\equiv}C-(CH_2)_2-C\substack{{\nearrow}O\\{\diagdown}O}$ (cholesteryl) (CPD5-10) (np)	K $\xrightarrow{80°C}$ I $\xleftarrow{69°}$ S_A $\xleftarrow{82°C}$
$CH_3-(CH_2)_3-C{\equiv}C-C{\equiv}C-(CH_2)_8-C\substack{{\nearrow}O\\{\diagdown}O}$ (cholesteryl) (CHD11-6) (np)	K $\xrightarrow{53°C}$ I $\xleftarrow{39°C}$ $S?$ $\xleftarrow{53°C}$
$Chl-O\substack{O{\diagdown}\\{\diagup}}C-(CH_2)_8-C{\equiv}C-C{\equiv}C-(CH_2)_8-C\substack{{\nearrow}O\\{\diagdown}O-Chl}$ (BCED11-11) (np)	K $\xrightarrow{114°C}$ I $\xleftarrow{76°C}$ S_A $\xleftarrow{114°C}$
$CH_3-(CH_2)_9-C{\equiv}C-C{\equiv}C-C\substack{{\nearrow}O\\{\diagdown}O}$—⬡—⬡—$O-CH_3$ (MBPD3-12) (p ?)[a]	K$\xrightarrow{54°C}$ N $\xleftarrow{86°C}$ I $\xleftarrow{28°C}$
$CH_3-(CH_2)_3-C{\equiv}C-C{\equiv}C-(CH_2)_8-C\substack{{\nearrow}O\\{\diagdown}O}$—⬡—⬡—$O-CH_3$ (MBHD11-6) (p +)	K $\xrightarrow{33°C}$ K' $\xrightarrow{72°C}$ I
$CH_3-(CH_2)_3-C{\equiv}C-C{\equiv}C-(CH_2)_8-C\substack{{\nearrow}O\\{\diagdown}O}-(CH_2)_2$—⬡—⬡—$O-CH_3$ (MEBHD11-6) (p +)	K $\xrightarrow{79°C}$ K' $\xrightarrow{86°C}$ I

Table I: (np) non-polymerizable monomer; (p +) polymerizable monomer under UV irradiation 254 nm, Chl: cholesteryl group; [a] *thermal decomposition.*

The three first monomers in this table, substituted with the cholesteryl moiety, present the liquid-crystalline order of smectic phases. These monomers do not polymerize in the solid-state under UV or γ irradiation. The biphenyl derivatives, except the MBPD3-12, do not present any liquid crystalline properties but are able to polymerize in solid-state under UV irradiation. The resulting polymerized crystals are red or red-orange, corresponding to the classical colour of conjugated chains of polydiacetylenes.

Two types of disubstituted diacetylenes can be distinguished from the results already reported on liquid crystalline derivatives (Schen 1988, Tsibouklis et al 1988): type I corresponds to a monomer in which the diacetylenic rod is in the mesogenic group, and type II corresponds to a monomer in which the diacetylenic rod is separated from the mesogenic group by a flexible spacer. In type I, liquid crystalline phase behavior may occur with the monomer and polymerization takes place to form nematic polymers (Tsibouklis et al 1988). In type II, polymerization in the mesophases is observed with rather small mesogen blocks attached to a short spacer, insuring a relatively rigid architecture of the molecule

(Schen 1988). In this work, the large cholesteryl block, which provides liquid crystalline properties, prevents monomer from polymerization (*CPD5-10, CHD11-6, BCED11-11*), whereas long spacers cannot be able to induce mesophases.

Three cholesteryl esters of propynoic, pentynoic and undecynoic acid (CP3, CP5, CU11) have been successivelly prepared. They present interesting liquid crystalline properties. Details are listed on table II.

(CP3)

H–C≡C–C

$K \xrightarrow{127\,°C} I$ $\xleftarrow[30\,°C]{} Ch \xleftarrow[105\,°C]{}$

(CP5)

H–C≡C–(CH$_2$)$_2$–C

$K \xrightarrow{107\,°C} Ch \xleftarrow{140\,°C} I$

(CU11)

H–C≡C–(CH$_2$)$_8$–C

$K \xrightarrow{51\,°C} S_A \xleftarrow{74\,°C} I$

Table II: acetylenic esters

Catalysts	Solvents	Temp. °C	Polym. time (h)	Monom. convers.%	Polym. %	Oligom. %
MoCl$_5$	dioxane	70	30	47	21	79
WCl$_6$	dioxane	30	24	75	87	13
MoCl$_5$/SnΦ$_4$	dioxane	30	24	71	21	79
WCl$_6$/SnΦ$_4$	dioxane	30	24	52	92	8
WCl$_6$	benzene	30	24	53	23	77
MoCl$_5$	benzene	30	24	54	43	57

Table III: polymerization of cholesteryl-pentynoate (CP5)

Polymerization of CP3, CP5, CU11 has been achieved in solution with transition metal halides (WCl$_6$, TaCl$_5$) with the addition of a cocatalyst, tetraphenyltin (SnΦ$_4$), generally used in polymerization of hindered acetylenes, with dioxane or benzene as solvent. The best results are obtained for CP5 and are listed in table III. The polymerization yield is given for the production of linear polymer and oligomer, the latter generally being a cyclic trimer.

CP5 gives with WCl$_6$ in dioxane a glassy red-polymer (pCP5) in good yield. The polymer analysis have been realized by gel permeation chromatography (GPC). The chromatogram presents two main peaks corresponding to the cyclic substituted trimer (M$_w$ = 1400) in low content and a linear cholesteryl polyacetylene in a proportion of 87%.

Optical characterization of cholesteryl polypentynoate:

The two components of the red product obtained with WCl_6 have been separated by reversed phase chromatography. The GPC chromatogram of each fractions shows a single peak. The molecular weight of the polymer fraction was evaluated from elution volume as a polystyrene equivalent. The obtained value, $M_w = 8 \times 10^4$, confirms that the number of repeat units of the polymer chain is > 100.

The spectroscopic properties of the pure polymer have been observed in solution or in solid-state thin film. A progressive loss of color of the solution takes place over a time of several days. This observation is confirmed in solution under vacuum in a sealed glass apparatus (Figure 1a) and in the solid thin film after a thermal treatment at about 80°C (Figure 1b).

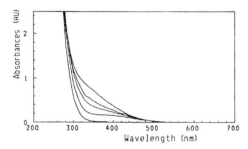
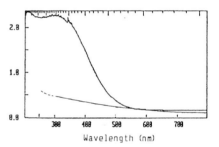

Figure 1a   Figure 1b

Such properties cannot be due to an oxidation of the double bonds of the polymer backbone and these spectroscopic changes must be due to a conjugation modification, i.e. a conformational change in the polymer backbone. This conformational change in solution and in solid state films corresponds to color evolution of the polymer from red to pale yellow.

From ^{13}C solid state NMR two peaks of the pCP5 spectrum are attributed to the two different carbons of the polymer backbone: the resonance at 136 ppm is attributed to $=C\!\!<$ and the next resonance at 118 ppm to the =CH- carbon atom. By comparison with the ^{13}C NMR signal given by Clarke et al (1982) on cis and trans $(CH)_x$, the initial conformation corresponding to the red glassy polymer is the cis structure.

This geometric structure of the backbone confirms the computer simulation of these cholesteryl substituted polyacetylenes. The trans conformation of the polyacetylenic chain is highly improbable owing to the steric hindrance of the first lateral methylene and also the lateral packing of the cholesteryl groups. The cis conformation gives more room for the lateral bulky substituent but a low steric hindrance remains between the first methylene of the substituent and the neighbouring ethylenic proton of the main chain (Figure 2). This polymer model is consistent with the spectroscopic properties of the polymer from the initial state: π-conjugation in initial cis geometry and time dependent conjugation loss.

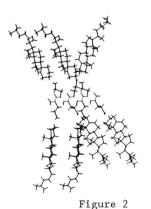

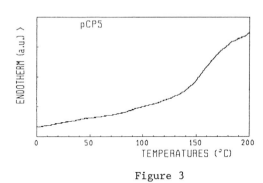

| Figure 2 | Figure 3 |

Liquid-crystalline properties

Liquid-crystalline properties have been studied on thin films of pCP5 in its red form by differential scanning calorimetry (DSC) and optical microscopy. The DSC endotherm, represented in Figure 3, is characteristic of a glass transition in the polymer. This curve does not present any apparent thermal transition peak during the first increase in temperature. The observation by microscopy reveals two optical transitions at 60° and 140°C. The thin film presents characteristic birefringent textures with a smectic order. These textures are under investigation by X ray scattering.

Non linear optical properties

The non linear optical properties (NLO) of the glassy film of poly-cholesteryl pentynoate were investigated by third harmonic generation (THG). The third order intensities, measured at 1.9 μm on the red and on the pale yellow film obtained by thermal treatment at about 80°C yield calculated cubic susceptibilities $\chi^{(3)}$ of about 10^{-13} esu. However, two main observations can be made:

First, the low value of $\chi^{(3)}$ is consistent with the low relative mass fraction of the polyacetylenic chain in the material. The calculated intrinsic cubic susceptibility is 20 times higher. It is comparable with the value found for high polarizable molecules such as CS_2. This modest value with respect to $\chi^{(3)}$ of the $(CH)_x$ chain, 10^{-10} esu, demonstrates the interest in the preparation of soluble substituted polyacetylenes with more conjugated backbone and with smaller substituents.

Second, the preparation of highly transparent thin films in the glassy state, from soluble polyacetylenes with more or less bulky substituents, appears possible.

3. EXPERIMENTAL

The synthesis of diacetylenic or acetylenic monomers has been described elsewhere (Le Moigne et al. 1989). Thermal properties were performed using a Mettler FP84 cell for DSC and microscopic observations. Substituted diacetylenes were polymerized by solid-state polymerization with γ-(^{60}Co) or UV-irradiation and polymerization of the cholesteryl acetylenic derivatives were obtained by reaction in solution with group 6 transition metal halides (Mo, W). The reaction is performed with the catalyst under vacuum in a sealed glass apparatus. The polymer was characterized by GPC standard methods. ^{13}C NMR spectra were recorded on a Brücker spectrometer at the magic angle spinning (MAS). Non-linear optical data have been obtained by third harmonic generation, the experimental arrangement have been previously described (Kajzar et al 1985).

4. CONCLUSION

We have been able to prepare conjugated polymers with mesogenic side-chain from substituted diacetylenes or terminal acetylenes. In this work we have presented the first characterization of a soluble polyacetylene with cholesteryl substituent. The initial polymer is obtained in a glassy red form corresponding to a cis planar conformation of the polymer backbone. The value of the cubic susceptibility $\chi^{(3)}$, is consistent with the relative low mass fraction of the polyacetylene chain in the material.

5. REFERENCES

Carter G M, Thakur M K, Chen Y J, Hryniewicz J V, 1985 Appl. Phys. Lett. **47**, 457
Clarke T C, McQuillan B W, Rabolt J F, Scott J C, Street J C, 1982 Mol. Cryst. Liq. Cryst. **83** 1
Garito A F, Teng C C, Wong K Y, Zammani'Khamiri O, 1984 Mol. Cryst. Liq . Cryst. **106** 219
Kajzar F, Etemad S, Baker G L, Messier J, 1987a Solid State Commun., **63** 1113
Kajzar F, Etemad S, Baker G L, J. Messier J, 1987b Synthetic Metals **17** 563
Kajzar F, Messier J, 1985c Thin Solid Films, **132** 11
Le Moigne J, Soldera A, Guillon D, Skoulios A, 1989 Liq. Cryst. in press
Milburn G H W 1988 Proc. *Nonlinear Optical Effects in Organic Polymers*, Ed. Messier J, Kajzar F, Prasad P, Ulrich D, NATO ASI series, Appl. Sci. **162** pp 149-158
Nunzi J M, Grec D, J. Appl. Phys. 1987, **62**, 2198
Ozcayir Y, Asrar J, Blumstein A, 1984 Mol. Cryst. Liq. Cryst. **110** 263
Ozcayir Y, Asrar J, Clough S B, Blumstein A, 1986a Mol. Cryst. Liq. Cryst. **138** 167
Ozcayir Y, Blumstein A, 1986b Mol. Cryst. Liq. Cryst. **135** 237
Sauteret C, Hermann J P, Frey R, Pradère F, Ducuing J, Baughman H, Chance R R, 1976 Phys. Rev. Lett. **36** 956
Schen M A, 1988 SPIE, Proc. *Nonlinear Polymers and Inorganic Crystals and Laser Media*, **971** pp 178-185
Tsibouklis J, Werninck A R, Shand A J, Milburn G H W, 1988 Liq. Cryst. **3** 1393

Inst. Phys. Conf. Ser. No 103: Section 2.5
Paper presented at Int. Conf. Materials for Non-linear and Electro-optics, Cambridge, 1989

Non-linear optical and electro-optical properties of polymer thin films for non-linear optical applications

J R Hill, P Pantelis and G J Davies,

British Telecom Research Laboratories, Ipswich, IP5 7RE, U.K.

ABSTRACT: Electrically oriented polymers are increasingly being recognized as potentially important materials for the construction of electro-optic devices utilizing second order nonlinear effects. These polymers can be divided into two distinct systems namely guest/host and single component routes. The origin of their optical nonlinearity, their relative stability and their potential applications are presented.

1. INTRODUCTION

Optical communications systems are already an important part of the telecommunications network and it is expected that their contribution will increase as advances in technology stimulate additional optical signal processing capabilities. The capacity of existing optical links could be increased far beyond their current use by developments such as wavelength and time division multiplexing and the introduction of coherent optical systems. In addition, the flexibility of optical networks could be greatly enhanced if optical signals could be switched directly from one route to another (space domain switching) without having to convert the optical signals to intermediate electronic ones, prior to their retransmission as light on another link. To enable such developments to take place, a variety of passive optical, optoelectronic and opto-optic components are desired.

To date, electro-optic devices have been fabricated in inorganic crystals such as semiconductors and lithium niobate $LiNbO_3$ (Nayar 1986) but recently, organic materials have attracted much attention as media in which to perform these operations. This interest arises from a combination of properties found in organic molecules possessing pi-electrons in delocalized molecular orbitals. The electronic properties of these compounds can be predicted by quantum mechanical calculations and their chemical structure can be readily altered using the expertise of synthetic chemists. The diversity of organic materials has introduced the concept of molecular engineering, whereby selected properties can be promoted to optimize the material's properties for each specific application. The optical nonlinearities are fast in these systems because the effects are derived almost exclusively from the movement of electrons, as opposed to large contributions from vibrational excitations as in lithium niobate (Garito and Singer 1982). In addition, because the electrons are generally confined within pi-orbitals which have little interaction between molecules, the properties of the bulk material can be conveniently considered to be the sum of the properties of the individual species (Hurst and Munn 1986). Organic molecules have also been reported by Kato (1980) and Velsko et al (1987) to have high optical damage thresholds which is desirable, bearing in mind the high power densities of optical fibre and laser systems.

In this paper we will concentrate on organic polymers for second order nonlinear optical applications where we wish to combine the versatility, low cost and processability of polymeric materials with the high intrinsic nonlinearity of organic compounds.

2. ORIGIN OF NONLINEAR OPTICAL EFFECTS

When high energy laser radiation propagates through matter, the intense electric field gives rise to a number of useful effects. These arise from nonlinearities in the polarization (dipole moment per unit volume) (P) induced by the radiation. This can be expressed as a power series in the field (E) (Equation 1) (Zernike and Midwinter 1973).

$$P = \epsilon_0 \chi^{(1)}E + \epsilon_0 \chi^{(2)}E^2 + \epsilon_0 \chi^{(3)}E^3 + ... \quad (1)$$

where: $\chi^{(i)}$ is the ith order susceptibility
and ϵ_0 is the permittivity of free space.

Particular properties may be considered as resulting from individual terms in χ. Thus the linear susceptibility $\chi^{(1)}$ may be considered responsible for the normal absorption and refractive index of the material. Nonlinear optical effects result from the second order susceptibility $\chi^{(2)}$ and these include optical frequency doubling, optical frequency mixing, optical rectification and the linear electro-optic effect. The susceptibilities are tensors and the induced polarization and electric field are both vectors. It can be demonstrated that in a medium which is centrosymmetric, all the even order susceptibilities are zero and the properties of the material depend only on the odd ordered terms. This symmetry requirement places a restriction on the packing of the molecules from which the material is comprised.

The third order term $\chi^{(3)}$ provides effects including optical frequency tripling and the quadratic electro-optic effect. The magnitude of the terms decreases with increasing order in E and the 4th and higher order terms are usually neglected. In part, this explains why (in spite of the requirement for noncentrosymmetric packing) greatest progress towards nonlinear optical devices has been towards those which depend on second order effects.

The linear electro-optic coefficients r_{ijk} and second-harmonic coefficients d_{ijk} are related to $\chi^{(2)}$ by: -

$$\chi^{(2)}{}_{ijk}(-w;w,0) = (n_{ii})^2{}_w.(n_{jj})^2{}_0.r_{ijk}(-w;w,0)/2$$

and $$\chi^{(2)}{}_{ijk}(-2w;w,w) = 2d_{ijk}(-2w;w,w)$$

where the tensorial nature of the coefficients is now explicit and n is the appropriate refractive index at the frequency shown (Garito et al 1983).

Because the nonlinear optical properties of a bulk organic solid can be described as a summation of the properties of the individual components, it follows that the nonlinear optical properties of these individual species can be described by a microscopic polarization (a change in dipole moment), also expressed in terms of powers of field (Equation 2).

$$p = \alpha E + \beta E^2 + \gamma E^3 + ... \quad (2)$$

Analogously to Equation 1, p and E are vectors and α, β and γ are tensors called respectively the linear polarizability and the second and third order hyperpolarizabilities. Values of α, β and γ can be related to $\chi^{(1)}$, $\chi^{(2)}$ and $\chi^{(3)}$ using the molecular packing density, local field factors at molecular sites, and a detailed structural or crystallographic knowledge of the bulk material (Oudar and Zyss 1982 and Zyss et al 1984).

Large values of β arise through the presence of an asymmetric mobile pi-electron system and low lying charge transfer states in a molecule. This can be provided by placing electron donor and acceptor groups onto a conjugated system (chromophore) so that polarization is easier in the direction from donor to acceptor than vice-versa. Considerable work has been done by Docherty et al (1985), Li et al (1986), Dirk et al (1986) and Chemla et al (1981) on the molecular origin of optical nonlinearities in organic materials and on the relationship between bulk and molecular properties.

3. POLYMERS FOR NONLINEAR OPTICS

Polymers offer many features which make them ideal materials for nonlinear optical devices (Le Barny 1987, Prasad and Ulrich 1988 and Heeger et al 1988). They are relatively inexpensive, and may be processed both by melt and solution spinning techniques. They can have good optical properties and, by chemical modification, their linear and nonlinear optical properties can be altered. They have adaptable electrical properties and are compatible with many semiconductor processing steps such as lithography, electroding and plasma etching.

For a polymer to exhibit second order nonlinear properties the bulk material must contain molecules possessing second order hyperpolarizability which are organised in such a way that there is no macroscopic centre of symmetry (Hurst and Munn 1986). To provide an even distribution of optically nonlinear molecules, a suitable species can be dissolved as a guest in an inert transparent polymeric host to form a solid solution (guest/host system) or the species can be chemically bound to the polymer. These molecules must then be induced to point, at least statistically, in a common direction. Provided that there is significant microscopic second order hyperpolarizability in the direction of the molecular ground state dipole moment of the guest, the required orientation may be induced by the application of an external electric field (Williams 1984). The molecules experience an energy minimum when they are aligned with their dipole moment in the field direction and, within the limits set by Boltzmann statistics, they take up this preferred orientation.

4. GUEST / HOST POLYMER SYSTEMS

In these systems, the guest is dissolved in the polymeric host and the natural centrosymmetry of the orientation of the guest molecules is broken by the application of an external DC electric field to the polymer, a process known as poling. In most guest/host mixtures the host polymer is in the glassy state at ambient temperatures. The almost complete absence of molecular motion of the polymer chains under these conditions prevents the guest from aligning within practical periods of time. On heating the mixture above the glass transition temperature (Tg), there is an increase in the free volume and segments of the polymer chains have greater mobility, allowing the guest to adopt a new orientational distribution in the presence of the external field. By cooling the mixture back under the Tg in the presence of the electric field, the new orientational distribution is retained for a time, which is dependent on the rate of thermal reorientation. In part, the rate of thermal relaxation will be a function of the exact cooling profile of the experiment because the free volume of the cooled polymer depends on the cooling rate (Billmeyer 1971).

The bulk nonlinearity of the whole system ($\chi^{(2)}$) is proportional to the degree of order and to the value of the microscopic second order nonlinear coefficient (β_{vect}) of the guest when vectored onto its ground state dipole moment (μ_g). The relationship between the degree of order, the dipole moment of the guest, the applied and local fields, and the temperature of the poling process has been calculated by Singer et al (1987). Usually, in both liquids and amorphous polymers, the order is found to be linear with the applied field and the dipole moment of the guest (Singer et al 1987 and Levine et al 1975). The nonlinear optical molecules in liquid crystalline polymers follow different statistical rules and can have up to five times more order than amorphous polymers under the same processing conditions (van der Vorst and Picken 1987). This advantage of increased order can be offset by the greater tendency to optical scatter in liquid crystals.

A number of host polymers have been investigated including poly(styrene) by Hampsch et al (1988) and poly(oxyethylene) by Watanabe et al (1988). Poly(methylmethacrylate) has been exhaustively researched as a host by Singer et al (1986), especially in combination with 4-(4'-nitrophenylazo)-N-ethyl-N-(2-hydroxyethyl)aniline (Disperse Red 1) as the guest. The second harmonic coefficient of this mixture was found to rise in proportion to the guest concentration. This is to be expected as the nonlinear properties of the film are the sum of the properties of the component species. A film containing about 12% by weight of the guest and aligned above the glass transition temperature of $110^{\circ}C$ at a poling field of 6.2×10^5 Vm^{-1} was found to have a second harmonic coefficient at $1.58\mu m$ of $d_{33} = 2.5 \times 10^{-12}$ mV^{-1}. However, as we, Singer et al (1988) and Hampsch et al (1988) have found, the poly(methylmethacrylate) based system is unstable and the second harmonic coefficient decays rapidly within a few days. Rapid decay was also observed with films employing poly(styrene) and poly(oxyethylene) as the host.

The system which we (Pantelis and Davies 1984, Hill et al 1987a and Pantelis et al 1988) have studied is unique, firstly because electrical ordering can be performed at room temperature by corona poling, and secondly because the orientation of the guest is maintained by an internal electric field. Room temperature corona poling has the advantage of easier processing conditions and permits the application of very high field strengths ($> 1 \times 10^7$ Vm^{-1}) without catastrophic electrical breakdown.

The host polymer that we selected was copoly(vinylidene fluoride/trifluoroethylene) (Figure 1a) which was prepared by Atochem UK from the corresponding monomers in the molar ratio 70:30. On a microscopic level the polymer exists as small crystallites surrounded by regions of polymer chains which are in an amorphous state and into which nonlinear optical guest molecules can be dissolved. When a corona discharge is applied across the guest/host mixture at room temperature, the guest molecules and the polymer's carbon-fluorine bonds (both within the amorphous region and individual crystallites) tend to align with the field. Those carbon-fluorine groups on the polymer chains comprising the crystalline regions retain their orientation when the field is removed. This permanent dipolar alignment of the polymer crystallites causes internal electric fields in the amorphous regions which are retained after poling (Hill et al 1987b) and which maintain the alignment of the optically nonlinear guest after poling. As a result, provided the guest remains in solution, the system has far greater stability than conventional thermopoled polymers.

(CF_2-CH_2)$_{0.7}$(CF_2-CFH)$_{0.3}$

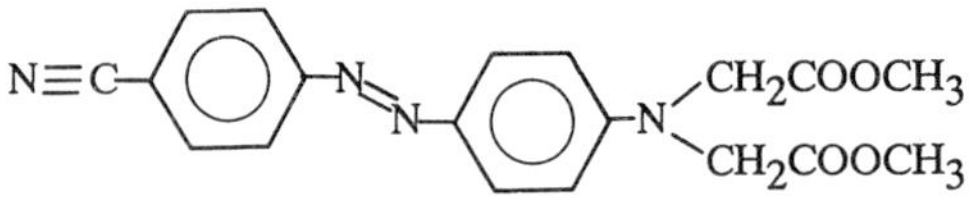

a) Foraflon 7030 host polymer   b)   4-(4'-cyanophenylazo)-NN-bis-
                                              -(methoxycarbonylmethyl)-aniline

Figure 1. BTRL guest/host system.

The guest we designed and synthesized for use with this host was 4-(4'-cyanophenylazo)-NN-bis-(methoxycarbonylmethyl)-aniline (Figure 1b). Crystals of this material are not suitable for second order nonlinear optical processes because of its centrosymmetric crystal structure. The guest material has a relatively high solubility in the host matrix and has both a large ground state dipole moment and microscopic second harmonic coefficient.

The second harmonic coefficients of the aligned guest/host films were determined by Maker fringe analysis (Jerphagnon and Kurtz 1970 and Singer et al 1986). In this technique, a plane parallel sample of the nonlinear optical material was rotated in the path of a fundamental beam of wavelength 1.064μm from a Q-switched Nd/YAG laser. Analysis of the intensity of the second harmonic light generated in the film as a function of incidence angle gives the second harmonic coefficient. Measurements were made relative to a Y-cut quartz plate which was taken as having $d_{11} = 4.6 \times 10^{-13}$ mV^{-1}. As predicted, the second harmonic coefficient of the poled mixture increased in proportion to the concentration of the guest (Figure 2). This was limited to a maximum of about 10% by weight, owing to precipitation of the solid guest from its host matrix. At this maximum concentration, the value of the second harmonic coefficient was found to be $d_{33} = 2.6 \times 10^{-12}$ mV^{-1} at 1.064μm. Inspection of Figure 2 shows that the second harmonic coefficient of the pure host is small and negative and that the coefficient of the mixture is dominated by the contribution of the optically nonlinear guest. As a result, the behaviour of the resulting mixture can be exactly controlled by changing either the concentration or the chemical structure of the guest.

The internal field in the polymer was measured by optical absorbance changes on poling (Hill et al 1987b) and has been found to remain stable over prolonged periods of time (Hill 1989). The main ageing mechanism appears to be precipitation of the guest but this can be prevented by selection of a suitable guest species. This is exemplified in Figure 3 which shows the room temperature ageing of the second harmonic coefficient over a period of a year for both cyanophenylazoaniline and aminonitrostilbene guests. The azo guest exhibited pronounced precipitation whereas the stilbene remained in solution and the nonlinear properties of this film did not change over this time.

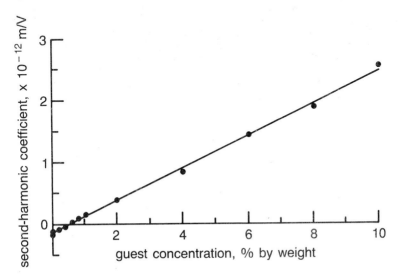

Figure 2. Second harmonic coefficient against concentration of guest

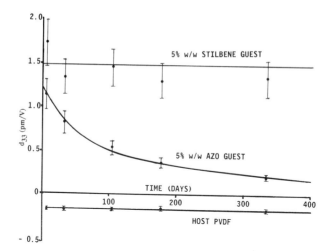

Figure 3. Room temperature ageing of second harmonic coefficient

5. SINGLE COMPONENT POLYMERS

The second harmonic coefficient of a guest/host system is ultimately limited by the concentration of the active guest material which can be incorporated into the polymer host. There has been increasing interest in covalently bonding the optically nonlinear species to the polymer to prevent precipitation and to maximize the number density of the nonlinear sites (see for example Esselin et al (1988) Singer et al (1988) Ye et al (1987) and Ye et al (1988)). Although polymers in which the optically nonlinear species forms an integral part of the polymeric chain have been suggested, the synthetic method more usually adopted is to produce a compound which contains the optically nonlinear species as a pendant group. This may be achieved by either chemically binding a nonlinear group onto an existing polymer or by polymerization of a monomer containing the nonlinear species. In order to decouple the motion of the nonlinear group from that of the polymer backbone it is usually distanced from it by a short flexible hydrocarbon chain. This produces a comb-like structure with the backbone forming the spine and the optically nonlinear species forming the teeth.

The spacer length greatly affects the physical properties of the final polymer (LeBarny et al 1986). If the spacer is long then the polymer is more likely to have liquid crystalline properties. For nonlinear optical applications isotropic polymers are usually desired owing to their generally lower light scatter and it is for this reason that we have concentrated our research on isotropic polymers. A series of acrylate polymers typified by Figure 4 have been custom synthesized for us by the Department of Chemistry at Lancaster University.

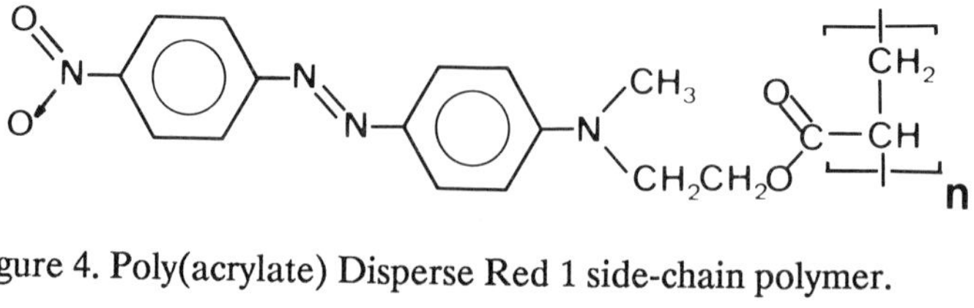

Figure 4. Poly(acrylate) Disperse Red 1 side-chain polymer.

As with amorphous guest/host solid solutions, when the polymer is heated above its T_g the pendant chromophores become mobile and can be ordered by the interaction of an applied electric field with their ground state dipole moment. In this system there is no internal electric field and the prime ageing mechanism is expected to be gradual loss of order with time. It has been observed by a number of workers including Esselin et al (1988), Singer et al (1988), Ye et al (1988) and Hill et al (1989), that the rate of thermal reorientation in side-chain polymers is much lower than in similar guest/host systems. However, it has yet to be conclusively demonstrated that thermal relaxation can be reduced to a level consistent with long term device stability.

Acrylate polymers and copolymers with Disperse Red 1 as the pendant group have been independently adopted by a number of research groups, including ourselves. Recently, alternative backbone polymers have been suggested including polyester and polystyrene (Ye et al 1987, Ye et al 1988 and DeMartino 1988). Attempts have also been made by Singer et al (1988), Ye et al (1988), DeMartino (1988) and DeMartino et al (1988), to modify the nonlinear properties by the introduction of other chromophores including nitroanilines, aminonitrostilbenes, nitrohydroxybiphenyls and 4-dicyanovinyl-4'(dialkyl)aminoazobenzene . However, because of the almost infinite variety of chemical structures which is possible with organic compounds, there still remains a large number of untested combinations of backbone polymers and nonlinear optical chromophores.

6. POLYMER OPTICAL WAVEGUIDES

One of the potential applications for optically nonlinear polymers is in the construction of active waveguides; the properties of which may be modified by the application of electrical or optical fields. Such active waveguides could be applied as thin films to almost any solid substrate and, because of the potentially high nonlinear coefficients of organic materials, could be physically short. Consequently, the possibility arises of integrating an active polymer waveguide overlayer and a semiconductor substrate containing all the required drive and control circuits to perform a number of operations on a single substrate. Alternatively, the polymer could used as an overlayer on half-coupler blocks or D-fibres to construct switches or filters.

We have investigated methods for the construction of slab waveguides equipped with the electrodes required for the application of both poling and modulation signals. A simple slab waveguide has been produced using processes previously developed for semiconductor resist technology and a typical processing sequence is given in Figure 5.

An aluminium layer was evaporated onto a substrate and sequentially layers of a low refractive index polymer, the high index active polymer and finally an upper layer of the same low index polymer were spun onto the substrate. A second aluminium film was evaporated onto the top layer to complete the sandwich and on this was spun a light sensitive etch resist. The resist was exposed via a shadow mask and then developed and etched to give a metallic stripe on the polymer surface. The metallic stripe formed a protective mask in a reactive ion etch allowing removal of unwanted polymer. In addition, the stripe acted as an electrical contact to the guide to allow poling and application of a modulation signal. A final step was the overcoating of the rib with an optically confining and protective polymer layer. This also allowed the guide end faces to be cut and polished. We have made polymeric waveguides of this kind and used them to demonstrate electro-optic behaviour in a Mach Zehnder interferometer.

Research into structures of this kind is currently being performed by a number of groups world-wide and different routes have been adopted to define the waveguide in the active polymer. Of particular note is a route pioneered jointly by The Celanese Research Company and Lockheed Research and Development Division whereby the guide is defined during the poling process. This process relies on the birefringence induced in the nonlinear polymer on poling. A metallic stripe is placed on a continuous film of the active polymer and the region under the stripe is thermopoled in the usual way. The index in the poled region is greater for one polarization of light than the index of the surrounding un-poled region and this difference confines the light locally. Electrically modulated Mach-Zehnder inteferometers, travelling wave modulators and directional couplers have been prepared in this way by Thackara et al (1988).

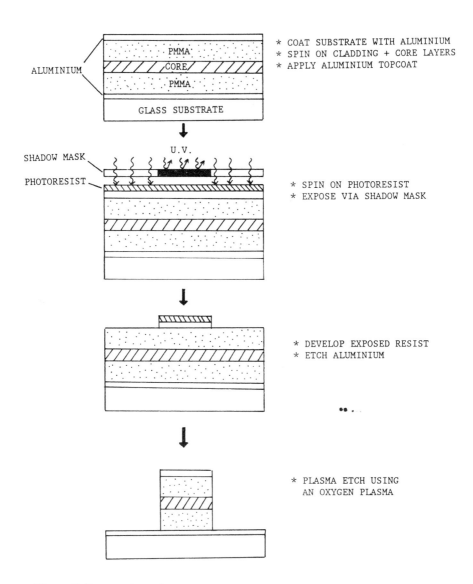

Figure 5. Preparation of a polymer waveguide by plasma etching.

7. DEMONSTRATOR POLYMER ELECTRO-OPTIC MODULATOR

Polymer based electro-optic devices are not restricted to waveguide geometries but can be formed into novel structures which could not have been constructed easily using either organic or inorganic single crystals. For example, one research group (Molmann et al 1988) has used poled polymers with nonlinear coefficients of the order of lithium niobate to construct a tunable Fabry-Perot etalon. We have made an alternative modulator (Hill et al 1988) which is relatively simple to construct and which forms a practical and effective demonstration of the linear electro-optic effect in polymers.

The device (Figure 6) was fabricated on a glass substrate which had interdigital indium-tin-oxide (ITO) electrodes defined on it by standard photolithographic and acid wet etch processes. On top of this electroded area was placed a quantity of the test polymer which was heated and spread under a cover slip to form a thin film. While the sample was held at a temperature just above the Tg, a DC potential was applied across the interdigital electrodes and this field was maintained while the sample was cooled through the Tg to room temperature. As a result of the electrically imposed ordering of the chromophores in the polymer, it developed a periodic change in refractive index which followed the pattern of the electrodes. This resulted in the formation of a diffraction grating in the polymer film which has remained stable over a period of almost 2 years.

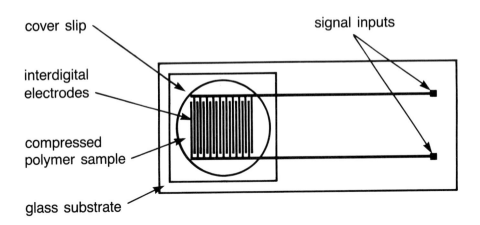

Figure 6. Schematic of a linear electro-optic modulator.

A Fraunhofer diffraction pattern was produced on passing a 0.633µm HeNe laser beam through the structure (Figure 7). The refractive index of the ordered regions could be electrically modulated via the linear electro-optic effect. Observations were made by selecting one of the diffracted spots with a photodiode and monitoring changes in diffraction efficiency when a signal was applied to the interdigital electrodes. In a test sample, the depth of modulation was linear with signal voltage with a maximum modulation depth of just over 17 percent.

We have attempted to approximately determine the magnitude of the electro-optic effect in the polymer by an analysis of the performance of the modulator relative to one which was constructed with nitrobenzene as the active medium and which operated via the DC Kerr effect. When thermopoled at 20 V/µm the polymer should achieve an electro-optic coefficient (r_{33}) of about $3 \times 10^{-11} \text{mV}^{-1}$ at 0.633µm. This value is approximately equal to the largest electro-optic coefficient of lithium niobate; although it should be noted that the close proximity of this laser source to the first electronic absorption band of the polymer will tend to resonance enhance the effect. At the longer wavelengths used in practical telecommunications systems the value would be lower and to achieve a coefficient of similar magnitude would require improved poling. Although our initial experiments have been performed at audio frequencies for experimental convenience, practical structures are expected to function at frequencies of many GHz.

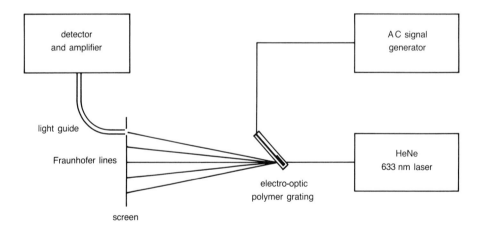

Figure 7. Schematic of the modulation experiment.

8. CONCLUSION

Organic polymers offer a unique and attractive area for nonlinear optical applications owing to their large nonlinear optical properties, fast response times and ease of processing. They have the additional advantages of low cost and compatibility with existing technologies. The techniques required for the preparation of both passive and active optical waveguides in polymers are under development. In the near future it should be possible to construct a range of devices, which are required by optical communications systems, from these materials. The ability of polymers to be shaped into a variety of physical forms will enable them to be exploited in a number of novel applications which are at present not readily accessible using existing materials.

9. ACKNOWLEDGEMENTS

We thank the Director of British Telecom Research Laboratories for permission to publish this paper. We would also like to thank our colleagues P L Dunn, D G Smith and J D Rush for their laser measurements, S N Oliver for chemical synthesis, C A Jones for waveguide preparation and R Heckingbottom and C R Day for useful discussions.

10. REFERENCES

Billmeyer F W 1971 'Textbook of polymer science' (NY London Sydney Toronto: Wiley-Interscience).

Chemla D, Oudar J L and Zyss J 1981 L'echo des Recherches (English issue) pp 47-60.

DeMartino 1988 'Condensation polymers with pendant side chains exhibiting nonlinear optical response' United States patent 4,757,130.

DeMartino, Choe E W, Khanarian G, Haas D, Leslie T, Nelson G, Stamatoff J, Stuetz D, Teng C C and Yoon H 1988 'Nonlinear optical and electroactive polymers' ed Prasad P N and Ulrich D R (NY:Plenum Publishing Corp) pp 169-187.

Dirk C W Twieg R J and Wagniere 1986 J Am Chem Soc 108 5387.

Docherty V J, Pugh D and Morley J O 1985 J Chem Soc Faraday Trans 2 81 1179.

Esselin S, LeBarny P, Robin P, Broussoux D, Dubois J C, Raffy J and Pocholle J P 1988 SPIE 971 120.

Garito A F and Singer K D 1982 Laser Focus Feb 59.

Garito A F, Singer K D and Teng C C 1983 'Nonlinear properties of organic and polymeric materials ACS Symposium Series 233', (Washington DC: American Chemical Society) pp 1-26.

Hurst M and Munn R W 1986 J Mol Elect 2 pp 35-41, 43-47, 101-105.

Hampsch H L, Yang J, Wong G K and Torkelson J M 1988 Macromolecules 21 526.

Hill J R, Dunn P L, Davies G J, Oliver S N, Pantelis P and Rush J D 1987 Elect Lett 23 700.

Hill J R, Pantelis P, Abbasi F and Hodge P 1988 J Appl Phys 64 2749.

Hill J R, Pantelis P and Davies G J 1987 Ferroelectrics 76 435.

Hill J R, Pantelis P, Dunn P L and Davies G J 1989 SPIE 1147-25.

Heeger A J, Orenstein J and Ulrich D R (Eds) 1988 'Nonlinear optical properties of polymers', (Pittsburgh,Pennsylvania:Materials Research Society).

Jerphagnon J and Kurtz S K 1970 Appl Phys 41 1667.

Kato K 1980 IEEE J Quantum Elect QE-16 1288.

Le Barny P 1987 Thin Solid Films 152 99.

Le Barny P, Ravaux G, Dubois J C, Parneix J P, Njeumo R, Legrand C and Levelut A M 1986 SPIE 682 56.

Levine B F and Bethea C G 1975 J Chem Phys 63 2666.

Li D, Ratner M A and Marks T J 1986 Chem Phys Lett 131 370.

Molmann G R, van der Vorst C P J M, Huijts R A and Wreesmann C T J 1988 SPIE 971 252.

Nayar B K and Booth R C 1986 Br Telecom Technol J 4(4) 5.

Oudar J L and Zyss J 1982 Phys Rev A 26 2016.

Pantelis P and Davies G J 1984 'Optical compositions' Patent GB 84/31682.

Pantelis P, Hill J R and Davies G J 1987 'Nonlinear optical and electroactive polymers' eds Prasad P N and Ulrich D R, (NY:Plenum Publishing Corp) pp 229-241.

Prasad P N and Ulrich D R (Eds) 1987 'Nonlinear optical and electroactive polymers', (NY:Plenum Publishing Corp) .

Singer K D, Kuzyk M G, Holland W R, Sohn J E and Lalama S J 1988 Appl Phys Lett 53 1800.

Singer K D, Kuzyk M G and Sohn J E 1987 J Opt Soc Am B 4 968.

Singer K D, Sohn J E and Lalama S J 1986 Appl Phys Lett 49 248.

Thackara J I, Lipscomb G F, Lytel R S and Ticknor A J 1988 'Nonlinear optical properties of polymers', eds Heeger A J, Orenstein J and Ulrich D R (Pittsburgh, Pennsylvania:Materials Research Society) pp 19-27.

Velsko S P, Davis L, Wang F, Monaco S and Eimerl D 1987 SPIE <u>824</u> 178.

van der Vorst and Picken S J 1987 SPIE <u>866</u> 99.

Watanabe T, Yoshinaga K, Fichou D and Miyata S 1988 J Chem Soc Chem Comm 250.

Williams D J 1984 Angew Chem Int Ed Eng <u>23</u> 690.

Ye C, Marks T J, Yang J and Wong G K 1987 Macromolecules <u>29</u> 2322.

Ye C, Minami N, Marks T J Yang J Wong G K 1988 Macromolecules <u>21</u> 2899.

Zernike F and Midwinter J 1973 'Applied nonlinear optics' (New York: John Wiley and Sons).

Zyss J, Nicoud J F and Coquillay M 1984 J Chem Phys <u>81</u> 4160.

Inst. Phys. Conf. Ser. No 103: Section 2.5
Paper presented at Int. Conf. Materials for Non-linear and Electro-optics, Cambridge, 1989

Polydiacetylene waveguide structures: thermal stability and optical losses in poly(4BCMU) films

Gregory L. Baker, J.A. Shelburne III, and Paul D. Townsend

Bellcore, 331 Newman Springs Road, Red Bank, NJ 07701-7040, USA.

ABSTRACT: High-quality low-loss films of the polydiacetylene poly(4BCMU) can be prepared by spin-casting techniques. Although such films are environmentally stable, they are unstable to thermal cycling. However, films annealed just below the disordering temperature (~ 115 °C), develop sharper optical absorption spectra, are thermally stable to nearly 100 °C, and based on preliminary waveguiding measurements, have lower optical losses. These results imply a maximum processing temperature of 100 °C for waveguide structures that incorporate poly(4BCMU) films.

1. INTRODUCTION

The large third order optical nonlinearities of conjugated polymers suggest their potential use in all-optical switching applications (1). Most proposed schemes for all-optical switching use the conjugated polymer in a waveguide format to gain confinement of the light and to maintain high optical intensities over macroscopic distances (2,3). While the active materials in future nonlinear optical devices must possess a large optical nonlinearity, they must also satisfy a set of important materials requirements to enable the fabrication of prototype devices. The development of such devices will require several steps: devising methods for the deposition of large-area high-quality films, the evaluation of their optical quality, the development of techniques for defining waveguide patterns in these films, and the testing of prototype nonlinear optical devices. Some time ago we embarked on a programme to develop generic solutions to these problems. Using poly(4BCMU) as the prototype nonlinear material, we have reported methods for depositing low-loss films (4,6), have characterised the optical losses in planar waveguide structures (4,5), and have developed two generic techniques for fabricating channel waveguide structures in nonlinear optical polymers; the lithographic definition of channel waveguide structures (6,7), and the formation of composite channel waveguides. (6,8) We note that from spectral measurements of third harmonic generation efficiency, poly(4BCMU) has a third order nonlinear susceptibility between 10^{-10} and 10^{-11} esu. (9)

Perhaps the most critical step in this sequence is the deposition of films of the nonlinear optical material on suitable substrates. A variety of techniques have been explored including dipping (10), spin-casting from solution (6), and vacuum evaporation (11). In this contribution, we focus on films prepared by spin-casting, the thermal stability of such films, and how thermal processing can improve the quality of poly(4BCMU) films.

2. SPIN-CAST FILMS OF POLY(4BCMU)

Poly(4BCMU) (Figure 1) belongs to a family of closely related, soluble polydiacetylenes. The common structural elements in these polymers are urethane substituents spaced from the polymer backbone by one or more methylene (CH_2) groups, and an n-butoxycarbonyl group attached to the urethane that provides enhanced solubility. Poly(4BCMU), having four methylene units, has been widely studied particularly in the context of its solvato- and thermochromic behaviour. (12-14) An important factor in these transitions is the ability of the urethane groups to form hydrogen bonds with either an adjacent urethane on the same chain, or alternatively with a urethane of a neighbouring chain. This ensemble of bonds forms a "hydrogen-bond lattice" with tie points between chains acting as crosslinks, and intramolecular hydrogen bonds favouring the formation of a planar ribbon-like polymer conformation.

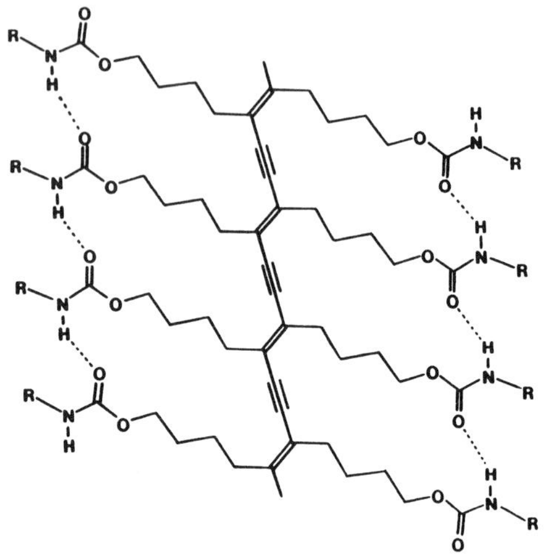

Figure 1. Chemical structure of poly(4BCMU)

Previously we reported that low-loss films of poly(4BCMU) could be spun from 2-4 wt.% solutions in cyclopentanone. (4,6) The films were evaluated at 1.06 μm by waveguiding experiments (4,5), and were found to have optical losses as low as 1dB/cm for the TM mode, and 5 dB/cm for the TE mode. As-spun films were optically anisotropic with a refractive index of 1.53 normal to the substrate surface, and an index of 1.60 parallel to the substrate. Anisotropy in spun films has previously been observed for polymers such as polystyrene, (15) and is caused by the spinning process. During spinning, a thin uniform layer of dissolved polymer is spread on the substrate, and as the solvent evaporates the film shrinks in thickness but not in area. Thus as the film thins, the once isotropic ensemble of polymer chains is "compressed" in the direction normal to the surface. Since solvent evaporation shrinks the film thickness at a rate faster than the relaxation times for the polymer, the dried film has a higher density of chains aligned in the plane of the film. For conjugated polymers, the π electrons of the conjugated backbone cause the refractive index to be large along the chain and smaller for directions perpendicular to the chain. Thus, the observed birefringence is much larger than is typically seen for saturated polymers.

Prism coupling experiments at 1.06 μm indicated that while spun films of poly(4BCMU) were stable toward environmental degradation, they were unstable at high laser-input powers. In these experiments, the resonant coupling angle is related to the refractive indices of the film, and varies as the effective indices for the film change. Increasing the intensity of low repetition-rate laser power prism-coupled into the film caused a change in the resonant coupling angle, but when the power was reduced to its initial value, the resonant coupling angle was not the original value, but one that corresponded to a decrease in the refractive index. Moving to a new spot on the film restored the coupling angle to its original value. We hypothesised that the small but finite optical absorption in these films combined with the strong 2-photon absorption seen in gels and films at 1.06 μm (16,17) resulted in sample heating, and that the polymer chains flowed under the pressure applied by the prism to yield a more isotropic film.

3.   THERMAL ANNEALING OF POLY(4BCMU) FILMS

To test this idea, we examined the thermal properties of spun films. Several prior studies (13,14) have considered the thermal properties of solution-cast films of poly(4BCMU), particularly the relationship between thermochromism and hydrogen bonding in the pendant substituents of the polymer. In general, Differential Scanning Calorimetry (DSC) experiments on cast films show an endothermic transition at ~ 115 °C (Figure 2) that corresponds to the melting of the hydrogen-bond lattice of the side-chains, with a second disordering transition appearing at ~ 140 °C. In contrast, the first endothermic transition for spun films is much broader, occurs at 109 °C, and extends to 30 °C.

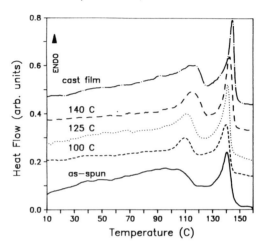

Figure 2. DSC traces for poly(4BCMU) films annealed at the indicated temperatures for 22h.

We suspected that this endothermic tail might be the cause of the thermal instability we had observed in the prism coupling experiments. Thermal transitions in polymers can often be sharpened by holding the polymer at temperatures just below the transition. Thus, we conducted a series of annealing experiments to see if the thermal instability could be overcome. In addition, we were interested in what effects the annealing might have on the optical properties of the polymer. For these experiments, spin-cast films (1 μm thick) were prepared under identical conditions, and after exhaustive pumping under dynamic vacuum to remove residual solvent, were isothermally aged at temperatures just above and below the disordering temperature. The films were examined by DSC at a heating rate of 10 °C/min under a flow of nitrogen. The data in Figure 2 show that the first disordering transition sharpened and shifted to higher temperatures as the annealing temperature was increased. A similar sharpening but smaller shift was seen for the second transition at 140 °C. Notable was the nearly flat baseline below 100 °C for samples annealed just below the transition temperature.

In terms of having the best thermal stability, a cursory examination of Figure 2 would suggest that annealing just below the second disordering transition at 140 °C would yield materials with the highest thermal stability. There are two complications, however. First, annealing at temperatures where the hydrogen-bond lattice of the side-chains is melted means that on cooling back through the transition, any disorder derived from the mobility of the polymer chains above the transition is locked into the sample. Thus, the two scans for samples held above the transition (140 °C and 125 °C in Figure 2) differ from the 100 °C scan by the weak tailing of the endotherm to lower temperatures. Based on the poor thermal stability of unannealed spin-cast samples, we concluded that even though the maxima of the first disordering transitions occur at higher temperatures for annealing at T> 135 °C, samples annealed at 100 °C will be more stable toward thermally induced changes over a wider operating range. A potential solution to this problem would be a two stage annealing scheme, the first stage at a high temperature to push the first disordering temperature to its maximum value, followed by a second annealing step at a temperature just below the transition to remove residual disorder in the film. Unfortunately such schemes result in films with degraded optical properties because of the growth at high temperatures of domains that efficiently scatter light.

4.   OPTICAL PROPERTIES OF ANNEALED POLY(4BCMU) FILMS

The optical properties of annealed thin films are presented in Figure 3. The absorption spectra for spun films differs from that of poly(4BCMU) gels (16) in that the exciton at the optical absorption edge in spun films is poorly defined. Thermal annealing below the disordering transition sharpens the exciton, but annealing above the transition causes it again to broaden. For annealing experiments above the transition (125 °C and 140 °C in Figure 3) the absorption showed no further change in shape but instead the absorption

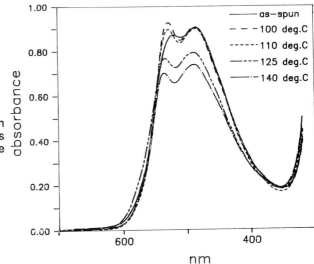

Figure 3. Optical absorption spectra of poly(4BCMU) films annealed for 22h at the indicated temperatures.

band uniformly reduced in intensity.   We note that the loss in intensity is inconsistent with chemical degradation of the polymer backbone, a process whose fingerprint is a blue-shift of the spectrum accompanying the loss in intensity.

The observation of domain formation in samples annealed well above the disordering transition suggests a simple, consistent explanation for our thermal and optical results, and provides guidelines for the thermal processing of poly(4BCMU) films.   Spin-cast poly(4BCMU) is highly disordered. The weakly defined exciton seen in the UV/vis spectra reflects disorder in the polymer backbone, (12) while IR (13,14) and DSC data describe a hydrogen-bond network in the side-chains that also is highly disordered. Annealing near the transition melts those bonds in less than optimum bonding arrangements, allowing the backbone conformation to relax into a more favoured (more planar) geometry with stronger hydrogen bonds between neighbouring urethane units along the chain. A sign of the increased order is the increased intensity of the exciton in the optical absorption spectrum. At temperatures above the disordering transition, the hydrogen-bond lattice is completely melted and the polymer chains can undergo self-diffusion and convert the initial birefringent film into an isotropic film. This transformation from a two-dimensional to three-dimensional arrangement of polymer chains can be followed by monitoring the optical absorption spectrum since the loss of order in the plane of the film causes a uniform reduction in oscillator strength. For the conversion of a pure two-dimensional array to an isotropic arrangement, the decrease in optical absorption is:

$$\frac{A_{3d}}{A_{2d}} = \frac{\int\limits_0^{2\pi} E \cdot E^* \cos^6\theta}{\int\limits_0^{2\pi} E \cdot E^* \cos^4\theta} = \frac{3/8}{1/2} = 0.75$$

Where E is the electric field vector and θ is the angle the vector makes with respect to the polymer backbone. We note that the spectrum (Figure 3) for the sample annealed at 140 °C is 80% of the intensity of the spectrum for the as-spun film, consistent with such an analysis.

Annealing at high temperatures for long times causes small domains to form that can be detected with an optical microscope. The appearance of these anisotropic domains is coincident with large optical losses in poly(4BCMU) and is suggests the first phases of a crystallisation process. Similar structures in poly(3BCMU) makes the spinning of low-loss films of poly(3BCMU) problematic.

5. SUMMARY

Understanding the materials properties of polymers designed for applications in nonlinear optics is crucial for the successful testing and evaluation of prototype nonlinear optical devices. We have shown for poly(4BCMU) how thermal instabilities are introduced in the film spinning process, and how they can be reduced by thermal annealing. By combining DSC and optical measurements, we determined that the thermal processing window for spun films of poly(4BCMU) extends to about 100 °C.

6. REFERENCES

1. For a general review, see: "Nonlinear Optical Properties of Organic Molecules and Crystals" Vol. 2; D.S. Chemla and J. Zyss eds., Academic Press, Inc., Orlando, FL USA, 1987.
2. R.K. Varsney, M.A. Nehme, R. Srivastava, and R.V. Ramaswamy *Appl. Opt.* **1986,** *25,* 3899.
3. C. Wachter, U. Langbein, F. Lederer *Appl. Phys. B* **1987,** *42,* 161.
4. P.D. Townsend, G.L. Baker, N.E. Schlotter, C.F. Klausner, S. Etemad *Appl. Phys. Lett.* **1988,** *53,* 1782.
5. P.D. Townsend, G.L. Baker, N.E. Schlotter, S. Etemad *Synth. Met.* **1989,** *28,* 633.
6. G.L. Baker, C.F. Klausner, J.A. Shelburne III, N.E. Schlotter, J.L. Jackel, P.D. Townsend, and S. Etemad *Synth. Met.* **1989,** *28,* 639.
7. G.L. Baker, C.F. Klausner *U.S. Patent 4,824,522*
8. N.E. Schlotter, J.L. Jackel, P.D. Townsend, and G.L. Baker *Appl. Phys. Lett.* (submitted).
9. W.S. Fann, private communication.
10. S. Mann, A.R. Oldroyd, D. Bloor, D.J. Ando, and P.J. Wells *Proc. SPIE* **1988,** *971,* 245.
11. T. Kanetake, K. Ishikawa, T. Koda, Y. Tokura, and K. Takeda Appl. Phys. Lett. **1987,** *51,* 1957.
12. R.R. Chance, G.N. Patel, and J.D. Witt *J. Chem. Phys.* **1979,** *71,* 206.
13. G. Walters, P. Painter, P. Ika, and H. Frisch *Macromolecules* **1986,** *19,* 888.
14. M.F. Rubner, D.J. Sandman, and C. Velazquez *Macromolecules* **1987,** *20,* 1296.
15. J.D. Swalen, M. Tacke, R. Santo, and J. Fischer *Opt. Commun.* **1978,** *18,* 387.
16. J.M. Nunzi, and D. Grec *J. Appl. Phys.* **1987,** 62, 2198.
17. P.D. Townsend, J.L. Jackel, G.L. Baker, J.A. Shelburne III, and S. Etemad *Appl. Phys. Lett.* (submitted).

Inst. Phys. Conf. Ser. No 103: Section 2.6
Paper presented at Int. Conf. Materials for Non-linear and Electro-optics, Cambridge, 1989

The relationship between structure and nonlinear optical properties in donor-acceptor polyenes

Charles W. Spangler[a], Tom J. Hall[a], Michelle L. Saindon[a], Robin D. Rogers[a], Ray K. McCoy[a], Robert R. Birge[b], Paul A. Fleitz[b] and Chian-Fan Zhang[b]

[a]Department of Chemistry, Northern Illinois University, DeKalb, Illinois 60115 USA; [b]Department of Chemistry, Syracuse University, Syracuse, New York 13244 USA

ABSTRACT: Second order hyperpolarizabilities have been determined for a series of donor-acceptor polyenes in which the D-A spatial relationship and the length of the pi-electron conduit has been varied, and ß values can now be related to substitution pattern and the number of double bonds in a polyenylic system. The absorption edge and maxima can also be controlled by altering the relative positioning of the donor-acceptor groups in the molecule and the identities of the D-A pairing. Crystal lattice packing can be affected in a similar fashion.

1. INTRODUCTION

In the past decade the increasing interest in the hyperpolarizabilities of organic molecules has resulted in several attempts to correlate second and third order effects with structure, hopefully leading to rational approaches to the design of molecules having specific and predictable NLO properties of eventual use in the construction of electro-optic devices. In general, it is now recognized that in order to maximize second order effects (ß), it is necessary to maximize the change in dipole moment between ground and excited states, and that the length of the pi-electron conduit in donor-acceptor type molecules is a critical factor in determining the magnitude of the effect. We would like to report some preliminary results of a large systematic study of these effects in a series of α,ω-diphenylpolyenes in which donor and acceptor substituents separated by varying length polyene chains have different relative positions on the aromatic rings. In this fashion it may be possible to gain control over the absorption characteristics of the molecules as well as the hyperpolarizabilities and thus explore both resonant and nonresonant behavior with available laser fundamental frequencies.

2. PREPARATION OF D-A POLYENES

Donor-acceptor-substituted diphenylpolyenes can be prepared by modified Wittig methodology as recently described by McCoy (1989).

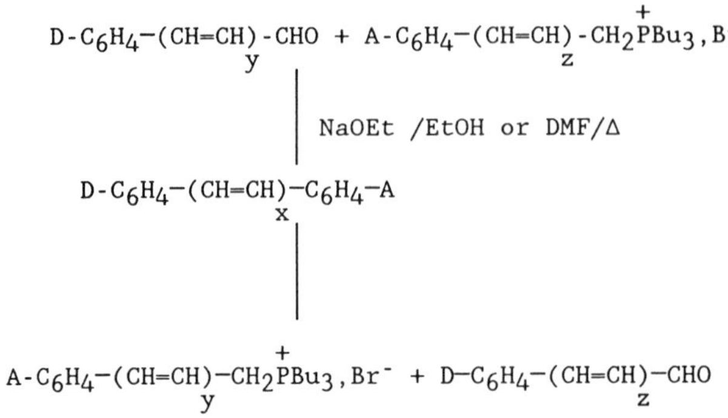

$$D-C_6H_4-(CH=CH)_y-CHO + A-C_6H_4-(CH=CH)_z-CH_2\overset{+}{P}Bu_3,Br^-$$

NaOEt /EtOH or DMF/Δ

$$D-C_6H_4-(CH=CH)_x-C_6H_4-A$$

$$A-C_6H_4-(CH=CH)_y-CH_2\overset{+}{P}Bu_3,Br^- + D-C_6H_4-(CH=CH)_z-CHO$$

D = Me$_2$N or MeO; x = y + z + 1

Figure 1.   Synthesis of D-A Diphenylpolyenes

Either donor or acceptor can be positioned on the phenyl ring in o, m or p positions, thus allowing several D,A spatial orientations for a given value of x. However since D and A groups can only communicate mesomerically when in o- or p-positions, we have limited the relative positioning in our study to p,p'; p,o'; o,p'; and o,o', as shown below.

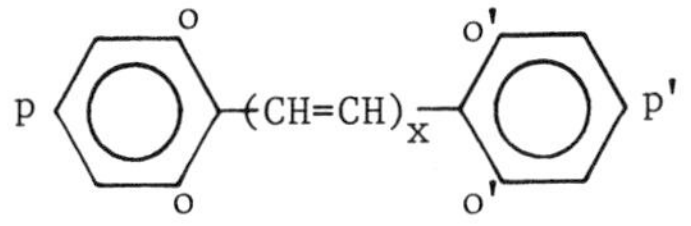

Figure 2.   Possible Substitution Patterns in D-A Diphenylpolyenes

3. ABSORPTION CHARACTERISTICS

In order to study nonresonant behavior, it is necessary that the molecule not absorb at the laser fundamental or harmonic. To a certain extent, the absorption maxima can be controlled by both the length of the pi-electron backbone, the relative strengths of the D-A pairings and the relative positionings of the substituents. The absorption characteristics of the molecules included in the present study are listed in Table 1.

Table 1.   Absorption Spectra for D-A Polyenes[a]

D	A	x	λ_{max}, nm	10^{-4}M threshold, nm
H	H	2	332	376
p-OMe	p'-NO$_2$	2	402	530
p-NMe$_2$	p'-NO$_2$	2	455	642
H	H	3	350	410
p-OMe	p'-NO$_2$	3	427	610
p-NMe$_2$	p'-NO$_2$	3	465	686
p-NMe$_2$	p'-NO$_2$	4	466	730
o-OMe	p'-NO$_2$	2	396	512
p-OMe	o'-NO$_2$	2	328	568
o-OMe	o'-NO$_2$	2	331	499

[a]In DMF solution

4. STRUCTURAL EFFECT ON ß VALUES

Leslie and coworkers (1987) have shown a direct relationship between ß values and pi-conduit length in a series of p-phenylene oligomers where D = p-NH$_2$ and A = p'-NO$_2$. Similarly Oudar and coworkers (1977a, 1977b) and Dulcic (1981) have both shown that variation of the acceptor and donor identities in the stilbene series (constant pi-length) can have a profound effect on both λ_{max} and μ_0ß. We have recently reported confirmation of these observations in a preliminary report (Spangler and coworkers, 1989). We would now like to extend these results to include the effect of D-A positioning in the molecule for a series with a fixed number of ene units (2).

ß values were determined by standard EFISH (electric field induced second harmonic) techniques at 1.06 μm in dioxane solution. These results are shown in Table 2.

Table 2. ß Values for D-A Polyenes

D	A	x	ß(X10^{-30}esu)[a]	λ_{max}[b], nm
p-MeO	p'-NO$_2$	2	156 ± 20	388
p-MeO	p'-NO$_2$	3	233 ± 35	406
p-NMe$_2$	p'-NO$_2$	2	508 ± 40	430
p-NMe$_2$	p'-NO$_2$	3	804 ± 61	449
o-MeO	p'-NO$_2$	2	114 ± 20	382
p-MeO	o'-NO$_2$	2	92 ± 12	325
o-MeO	o'-NO$_2$	2	73 ± 10	332

[a]10^{-4}M solution in p-dioxane [b]in p-dioxane solution

It can easily be seen that a length dependence for ß is confirmed in each series. However, when one compares the corresponding absorption characteristics for solutions of this concentration, it is clear that we are dealing with resonant behavior. Similarly we can see that D-A pairing effects on ß are strongly influenced by the strength of the donor (Me$_2$N > MeO). This has been confirmed most recently by Katz and coworkers (1988) for the "super" dithiolylidenemethyl donor. Similar effects for the super acceptor, tricyanovinyl, have also been found by this group.

We are intrigued by the possibilities of using a substitution pattern other than p-p'. As can be seen in Table 1, for a given polyene length, absorption characteristics are a function of D-A positioning. Thus, we can choose absorption characteristics in this fashion while maintaining ß at an adequate value for device consideration.

We were also surprised to find that variation of the D-A substitution pattern can affect the crystal packing. While most D-A polyenes preferentially crystallize in centrosymmetric structures, we found that the p-OMe, o-NO$_2$ diene crystallizes in a noncentrosymmetric pattern as shown in Figure 3. C$_{17}$H$_{15}$NO$_3$ crystallizes in the noncentrosymmetric space group Pna2$_1$ with a = 28.315(9), b = 24.853(9), c = 4.022(1)Å, P$_{calc}$ = 1.32 g/cm^3 for z = 8. The current R value of 0.087 was based on a least squares refinement of 765 independent observed [F$_0$ ≥ 3σ(F$_0$)] reflections. C$_{17}$H$_{15}$NO$_3$ crystallizes with two independent molecules in the asymmetric unit. The major difference in the two molecules involves the orientation of the methyl group: in the plane of the benzene ring

and oriented to the side of the molecule containing the nitro group is
one molecule and the side away from the nitro group in the other. Each
molecule bends slightly with the flattest dimension of both molecules
corresponding to the short c-axis of the unit cell. The nitro groups are
twisted out of the benzene ring plane by 37°. The two different forms
are illustrated in Figure 4.

5. CONCLUSIONS

It now appears possible that molecules can be designed for particular NLO
activity by varying the D-A pairing identities, the length of the pi-
electron system and the absorption characteristics dependence in D-A
positioning.

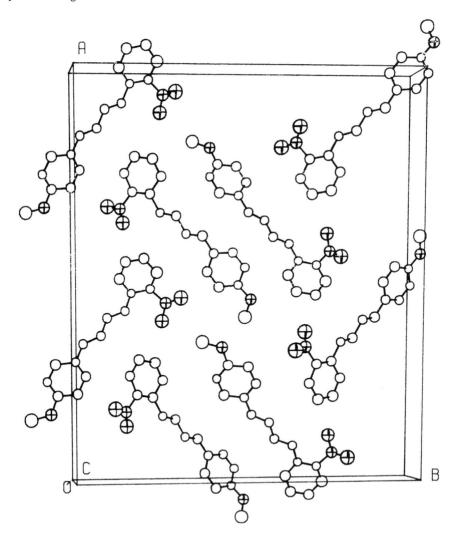

Figure 3.   Crystallographic Packing of 1-(p-methoxyphenyl)-4-(o-nitro-
            phenyl)-1,3-butadiene

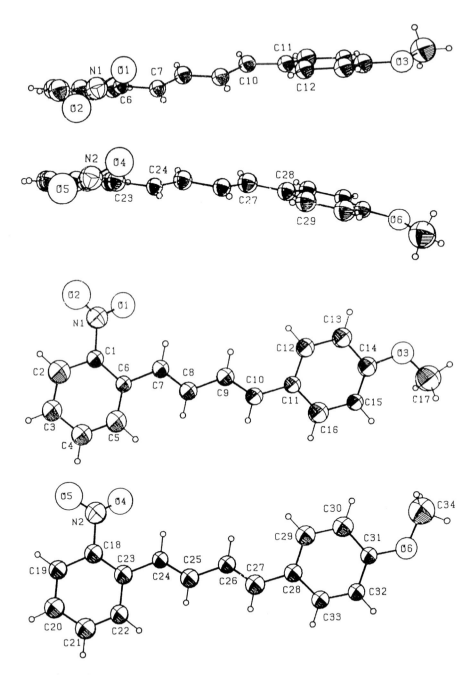

Figure 4. Alternate Views of the Individual Molecular Forms of
1-(p-methoxyphenyl)-4-(o-nitrophenyl)-1,3-butadiene

6.   ACKNOWLEDGEMENTS

The authors would like to acknowledge financial support of the Eppley
Foundation for Research, Inc.

7.   REFERENCES

Dulcic A, 1981 *J. Chem. Phys.* **74** 1559.

Katz H, Dirk C, Singer K and Sohn J 1988 *Mol. Cryst. Liq Cryst Inc.
Nonlin. Opt.* **157** 525.

Leslie T, DeMartino R, Choe E, Khanarian G, Hass D, Nelson G, Stamatoff
 J, Stuetz D, Teng C and Yoon H, 1987 *Mol Cryst Liq Cryst*, **153** 451.

McCoy R 1989 *Thesis* Northern Illinois University

Oudar J and Chemla D 1977a *J. Chem Phys* **66** 2664.

Oudar J 1977b *J. Chem Phys* **67** 446.

Spangler C, McCoy R, Birge R, Fleitz P, and Zhang C-F 1989 *Proc. Eng
Conf Molec Electronics* (In Press).

Inst. Phys. Conf. Ser. No 103: Section 2.6
Paper presented at Int. Conf. Materials for Non-linear and Electro-optics, Cambridge, 1989

239

Very large quadratic non-linearities of push-pull polyenes. Effect of the conjugation path and of the end groups

M Barzoukas[a], M Blanchard-Desce[b], D Josse[a], J-M Lehn[b] and J Zyss[a]

[a] Centre National d'Etudes des Télécommunications, 196 Avenue Henri Ravera, 92220 Bagneux, France
[b] Chimie des Interactions Moléculaires (ER 285 of the CNRS), Collège de France, 11 Place Marcelin Berthelot, 75005 Paris, France

ABSTRACT: The quadratic hyperpolarisability β of push-pull polyenes of increasing length and bearing various acceptor and donor groups has been studied by use of the Electric-Field-Induced-Second-Harmonic (EFISH) generation technique. The lengthening of the polyenic conjugation path results in a sharp increase of β with the number of double bonds n while substitution of a double bond by a triple bond in the polyenic linker leads to a reduction in the electronic polarisation. The benzodithia group is shown to be less effective as an electron donor than the dimethylanilino one, however this difference appears to fall-off as n increases. The same phenomenon is observed with electron withdrawing groups.

1. INTRODUCTION

Some organic molecules display quadratic hyperpolarisabilities leading to higher susceptibilities than those of most inorganic materials. It has been established that two major factors contribute to enhance significantly β:
- the asymmetry of the electronic distribution both in ground state and excited state configurations
- the length of the conjugated system
To induce the required asymmetric distortion of the electronic density, donor and acceptor substituents can be attached to opposite ends of a conjugated π system. The donor/acceptor interaction, through the conjugated system, gives rise to a low transition energy with a large oscillator strength. The amount of charge transfer (CT) from the donor to the acceptor is generally enhanced in this excited state relative to the ground state. Such molecules exhibit large quadratic hyperpolarisabilities and experimental studies - later accounted for by quantum chemistry calculations - relate the β enhancement to the existence of the CT transition (Nicoud *et al* 1987). A quantum two level model has been rightfully adopted as a simple method to evaluate β (Oudar 1977). An extension of the conjugation path between the donor and acceptor groups leads to a substantial increase in the value of β (Oudar 1977, Dulcic *et al* 1981, Barzoukas *et al* 1989). This was found to agree with the results of calculations on series of disubstituted molecules using semi-empirical methods (Zyss 1979, Morley *et al* 1987).

In this paper, the quadratic hyperpolarisability β of push-pull polyenes (Blanchard-Desce *et al* 1989) has been investigated by means of the EFISH technique. The experimental method and the results are gathered in Sec. 2. The molecules bear various donor and acceptor end groups, linked together by polyenic chains of different length (Figure 1). The donor groups are the following: benzodithia (compounds **1-5**), N,N-dimethylanilino (compounds **1'-5'**), julolidino (compounds **2"**). The chemical formulas of the acceptor groups are given in Figure 1: **a, b, c, d**. As shown in Sect. 3, there is a nearly quadratic dependence of β on the length of the conjugation path for series **1a-5a**. The efficiency of the substituents is seen to level-out at long chain lengths. An off-resonance ranking of the donor and the acceptor groups is established for a middle-sized conjugation path. Also the effect of the triple bond (molecule **3a**) is compared to that of the double bond (molecule **4a**).

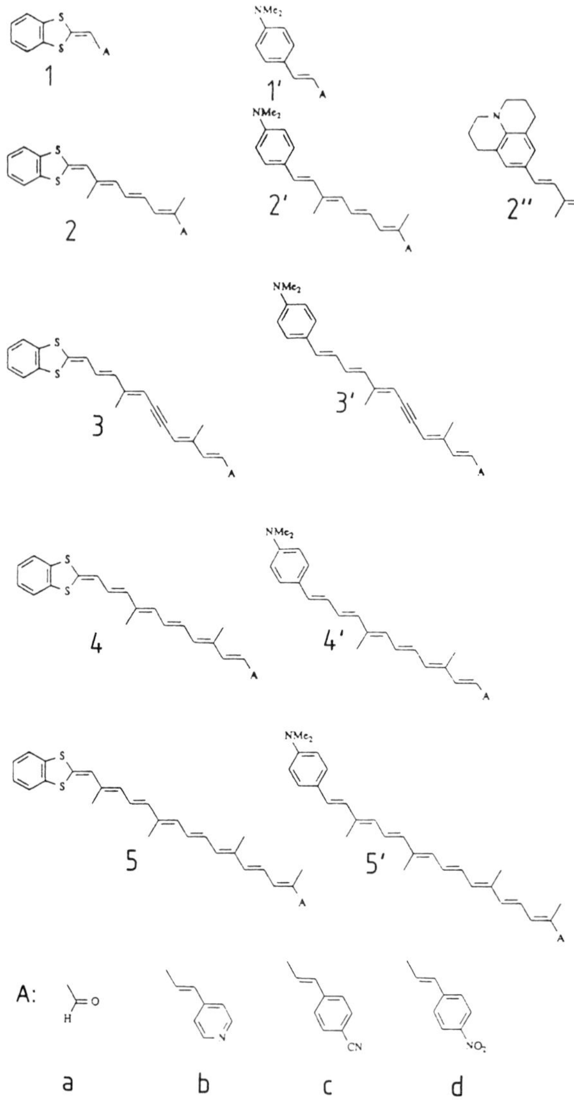

Fig. 1.   Structural formula of the push-pull polyene series.

2. EXPERIMENTAL DETAILS

The EFISH experiment is used to determine the molecular quadratic hyperpolarisabilities (Levine *et al* 1975, Oudar 1977). Solutions in chloroform, of increasing concentrations, are irradiated by a mode-locked Q-switched laser operating at 1.34 μm. The experimental procedure has been described extensively by Ledoux *et al* (1982). In all cases absorption in the sample is found negligible both at the fundamental (ω) and harmonic (2ω) frequencies. A mean microscopic hyperpolarisability γ^ρ of the solute molecule is derived from the measurement of the macroscopic susceptibility of the sample:

$$\gamma^\rho = \gamma(-2\omega;\omega,\omega,0) + \vec{\mu}.\vec{\beta}(-2\omega;\omega,\omega)/5kT \qquad (1)$$

The first term is the scalar part of the cubic hyperpolarisability tensor γ_{ijkl}, whereas the second term originates from the partial orientation of the permanent dipole moment $\vec{\mu}$ in the static field. $\beta(2\omega)$ - for $\beta(-2\omega;\omega,\omega)$ - is the vector part of the quadratic hyperpolarisability tensor:

$$\beta_i(2\omega) = \sum_j \beta_{ijj}(2\omega) \qquad (2)$$

The orientational contribution is usually assumed to be predominant over the electronic term γ^e (Oudar 1977, Barzoukas *et al* 1988); the scalar product $\vec{\mu}.\beta(2\omega)$ is then obtained directly from the EFISH measurement. The molecular dipole moments could not be measured experimentally. It can be suggested however - from the μ-values reported in the literature for similar push-pull compounds - that the dipole moments of the molecules studied in this work are relatively unaffected by the conjugation length (Hutchinson *et al* 1958, Mc Clellan 1963) and by the different donor/acceptor pairs (Mc Clellan 1963, Katz *et al* 1987), and can be estimated around 6-8 D. The variations of $\vec{\mu}.\beta(2\omega)$ will thus reflect mainly the dependence of $\beta(2\omega)$ on various parameters.

MOLECULE	λ_1	$\mu.\beta(2\omega)$	$\mu.\beta(0)$
1a	372	30	20
2a	456	1200	570
3a	466	2200	1000
4a	485	2700	1100
5a	500	7250	2800
1'a	384	320	200
2'a	450	2000	1000
3'a	461	4200	2000
5'a	498	8900	3400
2"a	480	2900	1200
2b	457	1500	715
1c	410	250	140
2c	465	1950	900
1d	452	1000	480
2d	488	2200	900

Table 1 : Experimental results. The maximum absorption wavelength λ_1 (in nm) is measured in chloroform. $\mu.\beta(2\omega)$ is determined at 1.34 μm in chloroform using the EFISH technique. The accuracy is 15%. All $\mu.\beta$ values are expressed in 10^{-48} e.s.u.

All experimental results are gathered in Table 1. As $\beta(2\omega)$ can be significantly affected by resonance enhancement, the relevant coefficient marking the intrinsic molecular properties is the static coefficient $\beta(0)$. The quantum two level model is used to describe molecules, taking into account only the predominating CT process that occurs between the donor and the acceptor groups. It follows that the β_{ijk} tensor is essentially one-dimensional along the CT (x) axis. For the molecules investigated in this work, x coincides with the $\vec{\mu}$ direction, so that finally $\beta(2\omega)$ is reduced to the CT component $\beta_{xxx}(2\omega)$. $\beta_{xxx}(2\omega)$ is related to the static hyperpolarisability coefficient by:

$$\beta_{xxx}(2\omega) = F(\omega,\omega_1) \ . \ \beta_{xxx}(0) \tag{3}$$

where $F(\omega,\omega_1)$ the dispersion factor away from resonance is given by:

$$F(\omega,\omega_1) = \omega_1^4/(\omega_1^2 - 4\omega^2)(\omega_1^2 - \omega^2) \tag{4}$$

here $h\omega_1$ stands for the transition energy to the first excited state, and ω is the frequency of the applied optical field.

The $\mu.\beta(0)$ values, derived from experimental data by use of the appropriate dispersion factor, are also listed in Table 1.

3. DISCUSSION

3.1 Effect of the Length of the Conjugation Path

For each series of molecules we have observed a red shift of the CT band together with a substantial increase of the quadratic hyperpolarisability, as the conjugation path lengthens. The off-resonance $\mu.\beta(0)$ values of molecules **5a** and **5'a** are exceptionally large. The estimated $\beta(0)$ hyperpolarisability - with $\mu{\sim}8D$ - amounts up to 50 times the $\beta(0)$ value of 4-nitroaniline.

The enhancement of $\beta(0)$ as a function of the number of double bonds is shown in Figure 2 for series **1a-5a** and **1'a-5'a**. The values of $\log\mu.\beta(0)$ are plotted against $\log n$, and the dependence of $\beta(0)$ on n for series **1a-5a** may be approximated by:

$$\beta(0) \propto n^{2.4}$$

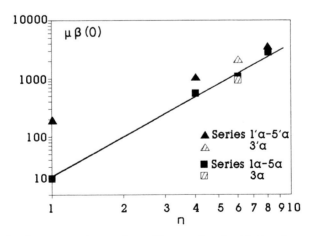

Fig. 2. Plot of $\mu.\beta(0)$ versus the number of intervening double bonds n, using logarithmic scales.

For series **1'a-5'a** the same trend is noted, although the value of molecule **1'a** falls out of line. It should be noted however that a phenyl ring has been arbitrarily included in the donor group, an approximation not valid for short chain lengths. These results are in good agreement with earlier experimental data reported by Dulcic *et al* (1981), where a nearly quadratic length dependence of β is shown. The dependence of $\beta(0)$ on n has also been investigated using semi-empirical calculations. Computed results given by Morley *et al* (1987) on two series of push-pull polyenes (series I: dimethylaminonitro polyenes and series II: dimethylamino polyenals) are consistent with present results. For series I the rise follows $n^{2.5}$ (n<10), and for series II the rise is in $n^{2.7}$. The present results do not agree with calculations reported by Li *et al* (1988) on series I, where an exponential dependence of $\beta(0)$ on n is indicated.

We have already shown that the rise in $\beta(0)$ is steeper for the series substituted with the less efficient donor group (Barzoukas *et al* 1989). It follows that the difference in efficiency between donor substituents reduces as the chain length increases. This result may now be extended to series substituted with different acceptor groups. This is illustrated by the $\mu.\beta(0)$ values obtained for molecules **1c-2c** and **1d-2d** (Table 1). The difference in donating (or accepting) power between donor (acceptor) groups falls off at long chain lengths. The efficiency of the donor→acceptor interaction (for different donor/acceptor pairs) appears to level-out as n increases.

3.2 Comparison between the Efficiency of Different Donor, Acceptor Substituents

The comparison between the efficiency of donor or acceptor groups should be carried out at a given chain length. Some of the donor and acceptor groups investigated here contain a conjugated part. So that for the same structure (**1, 2, 4** or **5**) but for different donor/acceptor combinations, the effective conjugation path differs. This is even more striking for the shortest compounds (struct. **1**), but becomes structurally less important for the longer ones. We have chosen to compare the static hyperpolarisabilities of middle-sized molecules (struct. **2**), since the effectiveness of donor or acceptor substituents levels-out as n increases. The following sequence of the donor (resp. acceptor) groups is established, by comparison between the off-resonance non-linearities of molecules **2a, 2'a, 2"a** (resp. **2a, 2b, 2c, 2d**):

The julolidino group is shown to be even more efficient than the N,N-dimethylanilino one. This agrees with studies of the effect of this group on the absorption spectra of push-pull olefins (Hassner *et al* 1984). The inclusion of the nitrogen atom in a cyclic structure rigidifies the conjugated system into an all-planar geometry, thus blocking in an optimised position the nitrogen lone pair. The resulting increase in the efficiency of the CT process is appreciable both in terms of maximum absorption wavelength and of $\mu.\beta(0)$. The efficiency of **b** as an electron withdrawing group has also been tested. It is seen to be more effective than the formyl (**a**) group. The efficiency of the donor/acceptor interaction for groups **c** and

d appears to level-out for the structure considered here. The $\mu.\beta(0)$ value of molecule **1d** is 4 times larger than that of molecule **1c**, whereas molecules **2c** and **2d** have comparable non-linearities. On the other hand the maximum absorption wavelengths are 23 nm apart.

3.3 Effect of the Triple Bond

Experimental results on molecules **3a** and **4a** show a decrease in the static hyperpolarisability, when a triple bond is included in the conjugation path instead of a double bond. The hypsochromic effect of the triple bond with regard to the double bond, and the drop in the quadratic non-linearity indicate a reduction in the electronic CT.

4. CONCLUSION

The quadratic hyperpolarisability of push-pull polyene systems increases rapidly with the number of double bonds ($\beta(0) \propto n^{2.4}$). The exceptionally large $\mu\beta(0)$ values of long polyene systems (n=8) evidence the potential of such compounds, which are likely candidates to be included in thin films such as Langmuir-Blodgett or poled-polymer films. The donating (withdrawing) strength of donor (acceptor) groups is seen to level-out with the lengthening of the conjugation path. Also the triple bond present in the conjugation path is shown to be a less efficient charge transmitter than the double bond. However the gain in transparency and stability of the molecular system could make up for the loss in non-linear efficiency.

The electronic cubic contribution to the EFISH non-linearity is negligible compared to the orientational contribution for 4-nitroaniline derivatives. This may not be true for longer conjugated chains, since the cubic hyperpolarisability $\gamma(-3\omega;\omega,\omega,\omega)$ of polyene systems is expected to rise as n^5 (Rustagi *et al* 1974). A combination of the EFISH and of the third harmonic generation techniques (related by a valid dispersion model) is presently being used to discriminate between the two contributions to the over-all EFISH non-linearity.

REFERENCES

Barzoukas M, Fremaux P, Josse D, Kajzar F and Zyss J 1988 Mat. Res. Soc. Symp. Proc. **109** pp 171-85

Barzoukas M, Blanchard-Desce M, Josse D, Lehn J-M and Zyss J 1989 Chem. Phys. **133** pp 323-9

Blanchard-Desce M, Ledoux I, Lehn J-M, and Zyss J 1989 Organic Materials for Non-linear Optics (Hann R A and Bloor D eds.) Publ. Roy. Soc. Chem. pp 170-5

Dulcic A, Flytzanis C, Tang C L, Pépin D, Fétizon M and Hoppilliard Y 1981 J. Chem. Phys. **74** pp 1559-63

Hassner A, Birnbaum D and Loew L M 1984 J. Org. Chem. **49** pp 2546-51

Hutchinson H M and Sutton L E 1958 J. Chem. Soc. pp 4382-6

Katz H E, Singer K D, Sohn J E, Dirk C W, King L A and Gordon H M 1987 J. Am. Chem. Soc. **109** pp 6561-3

Ledoux I and Zyss J 1982 Chem. Phys. **73** pp 203-13

Levine B F and Bethea C G 1975 J. Chem. Phys. **63** pp 2666-82

Li D, Ratner M A and Marks T J 1988 J. Am. Chem. Soc. **110** pp 1707-15

Morley J O, Docherty V J and Pugh D 1987 J. Chem. Soc. Perkin. Trans. II pp 1351-5

Nicoud J F and Twieg R J 1988 Non-linear Optical Properties of Organic Molecules (Academic Press New York: Chemla D S and Zyss J eds.) pp 227-96 and references therein

Mc Clellan A L 1963 Tables of Experimental Dipole Moments (Freeman, San Fransisco)

Oudar J L 1977 J. Chem. Phys. **67** pp 446-57

Oudar J L and Zyss J 1982 Phys Rev A **26** pp 2016-27

Rustagi K C and Ducuing J 1974 Opt. Com. **10** pp 258-61

Zyss J 1979 J. Chem. Phys. **71** pp 909-16

Optothermal nonlinearities in the polydiacetylene pTS

T G Harvey, W Ji, A K Kar, B S Wherrett

Department of Physics, Heriot-Watt University, Edinburgh EH14 4AS

D Bloor, P A Norman and D J Ando

Department of Physics, Queen Mary College, Mile End Road, London E1 4NS

ABSTRACT:　　Novel, low power nonlinear self-defocussing has been observed experimentally using the poly bis (p-toluene sulphonate) ester of 2,4-hexadiyne-1,6-diol (pTS). The strong anisotropy of the material's thermal conductivity, combined with the high observed thermo-optic coefficient, is shown to cause the phenomenon.

1.　INTRODUCTION

Self-defocussing (and self-focussing) Guha et al. (1985) of optical beams has been observed in many different media, all possessing an intensity-dependent contribution to their refractive index. The far-field diffraction pattern of an initially Gaussian cw laser beam which has traversed such a medium is normally characterised by a series of concentric rings (whose number increases with the laser power) surrounding a central maximum, Herman (1984). This pattern is the result of phase-changes introduced across the beam profile as it propagates through the nonlinear medium.

The diffraction pattern of the transmitted beam can be determined for any distance behind the medium by following the evolution of the beam amplitude and phase within the medium and then applying the Huygens-Fresnel propagation formalism to the electric field at the exit face Born & Wolf (1986). It has been assumed in previous studies however that the changes in phase and amplitude within the medium are cylindrically symmetric. This is the case for isotropic media.

In the polydiacetylene pTS, under CW conditions, the nonlinear contribution to the refractive index is thermal in origin and the anisotropy arises from anisotropic heat conduction parallel and perpendicular to the polymer chain direction [8]. Using this material, we have experimentally observed a far-field diffraction pattern which is significantly different to that which is normally observed. In particular, the transmitted beam breaks up into two beams in the far-field, with consequent implications to directional coupling. In the following we present a theory explaining and modelling this phenomenon.This theory is applicable to other materials and geometries.

2.   EXPERIMENT

The geometry of the experiment is shown in Figure 1. The incident Gaussian beam was polarised parallel to the polymer chain direction (y-direction) and propagated normal to the (1 0 0) crystal facet through a 137 μm thick single crystal pTS sample. The spot size of the incident beam was 25 μm (1/e^2 diameter) and wavelengths in the region of 700 nm were used. The transmitted intensity profiles for incident intensities in the range 0-1000 W cm^{-2} were recorded on a CCD camera linked to Hammamatsu frame store unit placed in the far-field behind the sample.

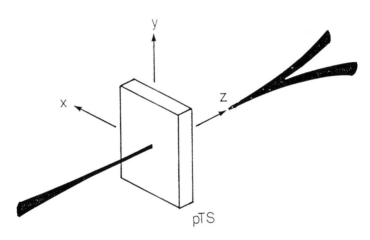

Fig. 1 Experimental geometry.

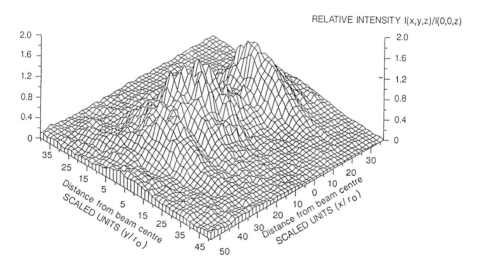

RELATIVE INTENSITY I(x,y,z)/I(0,0,z)

Fig. 2 Experimentally observed transmitted profiles at z = 10 mm (14 diffraction lengths) behind the pTS sample. Spot size is 25 μm. Incident irradiance = 964 W cm^{-2} (4.8 mW).

Figure 2 shows the experimental transmitted irradiance observed in the far-field behind the pTS sample for a high incident irradiance level. The profile exhibits preferential broadening into two beams or 'spots' and does not show the on-axis maximum characteristic of self-defocussed Gaussian beams. In modelling this phenomenon, the anisotropic distribution of the refractive index profile must be accounted for.

3. THEORY

The electric field at the back face of the sample can be expressed as a sum of Gaussian beams of decreasing spot size. Each component beam (or mode) within the summation can then be propagated to some distance, z, behind the sample using the theory of Kogelnik and Li (1966) and the resultant diffraction pattern calculated numerically. However, in the case where the nonlinear refractive index profile within the medium is not Gaussian but instead, due to some anisotropy, can be better represented by an elliptical Gaussian form (i.e. $\exp[-(x^2/r_x^2 + y^2/r_y^2)])$, the electric field at the exit plane of the sample would now become:

$$E(x,y,o) = E(o,o,o) \left\{ \exp\left[-\left(\frac{x^2+y^2}{r_0^2}\right)\right] + iX \exp\left[-\left(\frac{2x^2}{r_x^2} + \frac{2y^2}{r_y^2}\right)\right]\right\}, \quad (1)$$

where

$$X = \frac{\omega}{c} \int_0^L n_2 I_o(z) \, dz,$$

r_0 is the $1/e^2$ radius of the incident Gaussian beam, r_x and r_y are the radii of the induced elliptical Gaussian refractive index profile. I_0 is the irradiance at the beam centre in the medium and n_2 the refractive index nonlinear coefficient. This electric field can be expressed as a sum of elliptical Gaussian beams of decreasing radii $W_{mj}^2(o)$:

$$E(x,y,o) = E(o,o,o) \sum_{m=0}^{\infty} (iX)^m \exp\left[-\left(\frac{x^2}{W_{mx}^2(o)} + \frac{y^2}{W_{my}^2(o)}\right)\right], \quad (2)$$

where now, for j=x or y :

$$W_{mj}^2(o) = \frac{r_j^2 r_o^2}{(2mr_0^2 + r_j^2)}$$

The propagation of each elliptical Gaussian mode of specific m-value, is described by Yariv (1985). At a distance z behind the sample, the electric field is :

$$E(x,y,z) = E(o,o,o) \sum_{m=0}^{\infty} \frac{(iX)^m}{m!} (1 + (z^2/d_{mx}^2))^{-1/4} (1 + (z^2/d_{my}^2))^{-1/4}$$

$$\times \exp\left[- iP_m(z) - x^2\left(\frac{1}{W_{mx}^2(z)} + \frac{ik}{2R_{mx}(z)}\right) - y^2\left(\frac{1}{W_{my}^2(z)} + \frac{ik}{2R_{my}(z)}\right)\right]. \quad (3)$$

In the above, $k = \frac{\omega}{c}$,

$$W_{mj}^2(z) = W_{mj}^2(o) (1 + (z/d_{mj})^2) , \quad d_{mj} = \frac{kW_{mj}^2(o)}{2} ,$$

$$R_{mj}(z) = j (1 + (d_{mj}/z)^2) ,$$

$$P_m(z) = -1/2 \tan^{-1} (z/d_{mx}) - 1/2 \tan^{-1} (z/d_{my})$$

RELATIVE INTENSITY I(X,Y,z)/I(0,0,z)

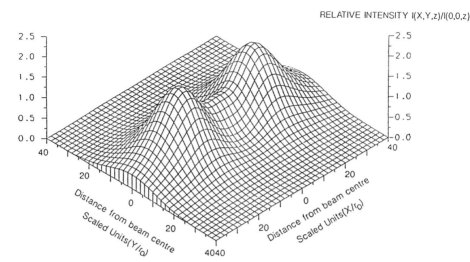

<u>Fig. 3</u>Far-field intensity distribution as calculated from equation (3) for the parameters $r_x = r_o$, $r_y = 2r_o$, X(total on-axis phase change) = 4 and z = 14 diffraction lengths.

Figure 3 shows an example solution to (3) for the particular case $r_x = r_o$, $r_y = 2r_o$, X = 4 and z = 14 diffraction lengths. Solution of the anisotropic heat equation with the ratio of thermal conductivities parallel and perpendicular to the polymer chain direction, $k_{\parallel}/k_{\perp} = 10$ at room temperature (Fig. 4) Morelli et al. (1986), Carlsaw & Jaeger (1959), has shown that these values are comparable to the anisotropy (defined as the ratio $\frac{r_y}{r_x}$ for an incident Gaussian beam) of the refractive index distribution induced in pTS in our experiments. The two-beam break-up is clearly predicted. Significantly also, the two-peaked structure propogates to long distance beyond the sample. Figure 4 demonstrates the maintainance of the two-beam propogtion for different amounts of anisotropy, showing the increase in the peak separation with the latter. This calculation was carried out for index profiles of minor axis radius equal to that of the incident irradiance (r_o) and varying major axis values. If the minor axis is greater than r_o then the resulting separations are smaller for any given ellipticity , i.e. the beam experiences effectively less anisotropy. This is the reason for the difference in contrast between figs 2 and 3.

One application of defocussing is in the direct measurement of the thermo-optic coefficient $\delta n/\delta T$ (or the nonlinear refractive n_2 in the case of an electronic nonlinearity) Weaire et al. (1981). For pTS we have experimentally measured $\delta n/\delta T$ in the band tail region 690-720 nm.It is found to be approximately constant at a value of $\sim - 1 \times 10^{-3}$ K^{-1}. It is as a result of this large value that the beam-splitting behaviour can be observed at power levels of only a few milliwatts.

In conclusion therefore, we have observed a new type of low-power nonlinear self-defocussing using the polydiactylene pTS. The phenomenon is the result of anisotropic heat conduction in the material combined with a large thermo optic coefficient.The theoretical model presented predicts this defocussing behavior and is applicable to other materials and geometries. For particular applications, in which the central beam irradiance might be restriced at a certain level or where efficient two-beam splitting is required, there is an optimum choice of system. The material choice would be to achieve maximum anisotropy but for minimum diffusion of the refractive index profile with respect to the incident beam radius. Detection at different distances beyond the sample may be used to control the level of power limiting; for beam-splitting the trade-off between insertion loss and contrast determines the optimum detector positions.

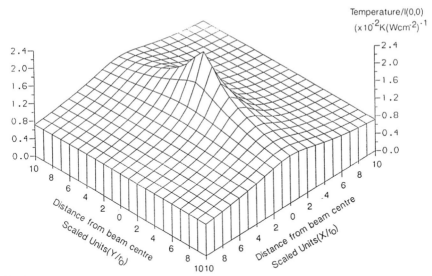

Fig. 4. Transverse temperature distribution (scaled to the peak incident irradiance) at the front face of the pTS sample for a 25 μm spot-size, $\alpha = 264$ cm^{-1} and $k_\perp/k_{\parallel} = 10$.

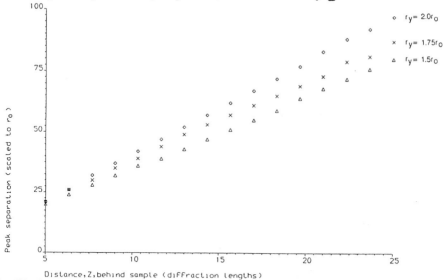

Fig. 5 Peak separation as a function of distance behind the sample for increasing anisotropy of the nonlinear refractive index distribution. $r_y = 1.5\ r_0$, $r_y = 1.75\ r_0$, $r_y = 2.0\ r_0$ Beam-splitting, which is enhanced with increasing anisotropy is shown.

4.   ACKNOWLEDGEMENTS
SERC support through the JOERS program is acknowledged. T G Harvey is grateful for financial support received from ICI.

5.   REFERENCES

Born E and Wolf E 1986 "Principles of Optics", 6th Ed. (Pergamon, Oxford), p. 428.
Carslaw H S and Jaeger J C 1959 "Conduction of Heat in Solids", 2nd Ed., Oxford
    University Press, Chapter 1.
Guha S, Van Stryland E W and Soileau M J 1985 Opt. Lett., **10**, 285.
Herman J A 1984 J. Opt. Soc. Am. B, **1**, 729.
H. Kogelnik and T. Li, Appl. Opt., **5**, 1550 (1966).
Morelli D T, Heremans J, Sakumoto M and Uher C 1986 Phys. Rev. Lett., **57** (7),
    869.
Weaire D, Wherrett B S, Miller D A B and Smith S D, 1981 Opt. Lett., **4**, 331.
Yariv A, "Optical Electronics" 1985 3rd Ed., Pub. J. Wiley, N.Y., p. 47.

Author Index